Automated Vehicles: Safety of the Intended Functionality (SOTIF)

Automated Vehicles: Safety of the Intended Functionality (SOTIF)

JUAN R. PIMENTEL

Professor of Computer Engineering
Kettering University

SAE INTERNATIONAL®

Warrendale, Pennsylvania, USA

400 Commonwealth Drive
Warrendale, PA 15096-0001 USA
E-mail: CustomerService@sae.org
Phone: 877-606-7323 (inside USA and Canada)
Fax: 776-4970 (outside USA)

Library of Congress Catalog Number 2019931391
SAE Order Number PT-205
http://dx.doi.org/10.4271/pt-205

Information contained in this work has been obtained by SAE International from sources believed to be reliable. However, neither SAE International nor its authors guarantee the accuracy or completeness of any information published herein and neither SAE International nor its authors shall be responsible for any errors, omissions, or damages arising out of use of this information. This work is published with the understanding that SAE International and its authors are supplying information, but are not attempting to render engineering or other professional services. If such services are required, the assistance of an appropriate professional should be sought.

ISBN-Print 978-0-7680-0235-5
ISBN-MediaTech 978-0-7680-0238-6
ISBN-prc 978-0-7680-0244-7
ISBN-epub 978-0-7680-0247-8
ISBN-HTML 978-0-7680-0268-3

To purchase bulk quantities, please contact: SAE Customer Service

E-mail: CustomerService@sae.org
Phone: 877-606-7323 (inside USA and Canada)
Fax: 776-4970 (outside USA)

Visit the SAE International Bookstore at books.sae.org

contents

CHAPTER 3

The Development of Safety Cases for an Autonomous Vehicle: A Comparative Study on Different Methods   27

CHAPTER 4

Autonomous Vehicle Sensor Suite Data with Ground Truth Trajectories for Algorithm Development and Evaluation   39

CHAPTER 5

Integrating STPA into ISO 26262 Process for Requirement Development — 53

CHAPTER 8

Bayesian Test Design for Reliability Assessments of Safety-Relevant Environment Sensors Considering Dependent Failures   101

CHAPTER 9

Challenges in Autonomous Vehicle Testing and Validation   125

Introduction

Safety has been ranked as the number one concern for the acceptance and adoption of automated vehicles and understandably so since safety has some of the most complex requirements in the development of self-driving vehicles. The recent fatal accidents involving self-driving vehicles have made it clear that safety is paramount to the acceptance, testing, verification, and deployment of automated vehicles. In conventional automotive systems, safety hazards are due to hardware (HW) or software (SW) faults or errors. However, automated vehicles are highly sensitive to safety hazards due to performance limitations rather than traditional HW or SW faults or errors. For example, it is well known that spurious and missed detections in bad weather pose a significant risk to an automated vehicle. Safety of the intended functionality (SOTIF) is a safety category that addresses hazards due to performance limitations of an autonomous vehicle, particularly those of the perception system. This book will include ten papers that will help in characterizing the safety of automated vehicles when the hazards are due to performance limitations rather than driving error or faults or errors in their sub-components.

Index Terms

Automated vehicles

Self-driving vehicles

Autonomous vehicles

Safety

SOTIF

Multi-agent safety

Functional safety

Safety measurement

Safety evaluation

I.1 **Introduction**

Automated vehicles (AVs), also called autonomous or self-driving vehicles, have the potential to reduce accidents, help with the environment, reduce congestion, help the elderly and other disadvantaged populations, and produce other societal benefits [1, 2, 3, 4, 5]. However, the touted advantages of autonomous vehicles and those including latest advanced driver assistance system (ADAS) features are turning out difficult to sell to the public than many manufacturers and tier 1s have anticipated. Much of the early euphoria of self-driving vehicles is diminishing in the wake of some recent accidents involving AVs with varying degrees of automation. A recent online marketplace for buying and selling cars found 69% of respondents are scared of autonomous automobiles. It is also found that these people found technology in cars helpful (58%), but only 12% said ADAS and infotainment features were a "must-have." The survey asked more than 1,000 respondents from across the United States geographically and across age groups, although it should be noted the biggest group was 60+ years old*. Accidents involving self-driving vehicles are inevitable; as it is the case with other industries, accidents have happened and will happen no matter the efforts made to avoid them. The National Transportation Safety Board (NTSB) has issued a report† on a Tesla accident on May 7, 2016, and two preliminary reports on an Uber accident‡ on March 18, 2018, and a Tesla accident§ on March 23, 2018. After analyzing these reports, what is disturbing are the details associated with these accidents which indicate that as an industry, we may need to go back to safety 101¶.

Regarding the aforementioned accidents, it would not be so bad if the safety systems of the vehicles in question were designed and functioning properly according to their stated automation level. In the case of the Tesla accident in Florida, the vehicle failed to activate the forward collision warning (FCW) system and automatic emergency braking (AEB). In the case of the Uber accident, emergency braking maneuvers were not enabled while the vehicle was under computer control, to reduce the potential for erratic vehicle behavior, and the system relied on the vehicle operator for safety. In the Tesla accident in California, the vehicle failed to detect a damaged crash attenuator and hit it at a speed of about 71 mph.

After recent incidents and mishaps involving AVs such as those described above, it is clear that there is much room for improvement not only by

---

* http://analysis.tu-auto.com/autonomous-car/shifting-public-acceptance-autonomous-tech?NL=TU-001&Issue=TU-001_20180723_TU-001_235&sfvc4enews=42&cl=artic le_2_2.
† https://www.ntsb.gov/investigations/AccidentReports/Reports/HAR1702.pdf.
‡ https://www.ntsb.gov/investigations/AccidentReports/Reports/HWY18MH010-prelim. pdf.
§ https://ntsb.gov/investigations/AccidentReports/Reports/HWY18FH011-preliminary. pdf.
¶ https://www.eetimes.com/document.asp?doc_id=1333446&_mc=RSS_EET_EDT&utm_ source=newsletter&utm_campaign=link&utm_medium=EETimesWeekInRevi ew-20180721.

manufacturers but also by government regulations, researchers, the general public, and other stakeholders. Over the past few months, the media has been full of headlines such as "How Safe Is Driverless Car Technology, Really?", "Autonomous Cars: How Safe Is Safe Enough?", and "How safe should we expect self-driving cars to be?" In addition, some industry analysts and safety experts are offering advice to tech and automotive companies to reconsider their safety programs. There is also some agreement that "the self-driving car industry's reputation has suffered a setback," and the question is how to fix it*. It appears that AV companies are much more stringent when using semiconductor devices and EDA tools demanding that they conform to ISO 26262 than using the same yardstick for their own safety-critical designs.

So what is there to do? Safety is not new; at least for the last 60 years, it has been successfully applied in several industries such as nuclear, avionics, process control, automotive, and others. What is unique and special about the safety of self-driving vehicles? What should be the emphasis for a more effective AV safety program? What are the roles of governments, standards, testing, verification, validation, and sound safety engineering efforts? Addressing issues regarding autonomous vehicle safety is challenging [7]. Currently, as an industry, we just do not fully understand the nature of self-driving vehicle safety and how to design safe AVs. For example, there is little discussion on ways to estimate, analyze, compute, or measure the level of safety of an AV design or AVs. We need to begin by fully characterizing it, and this book series is an effort in this direction.

Some manufacturers such as Waymo cite their recent milestone of 8 million miles driven on public roads as a measure of the safety achieved by their self-driving vehicles†. However, it is not clear how a certain number of millions of miles driven contribute to the safety level of self-driving vehicles. Some industry analysts believe that policy makers and city officials overseeing infrastructure will be the most important players in reshaping the self-driving vehicle safety landscape. For example, the National Highway Traffic Safety Administration (NHTSA) has issued a voluntary guidance whose purpose is to help designers of automated driving systems (ADSs) analyze, identify, and resolve safety considerations prior to deployment using their own, industry, and other best practices‡. It outlines 12 safety elements, which the agency believes represent the consensus across the industry, that are generally considered to be the most salient design aspects to consider and address when developing, testing, and deploying ADSs on public roadways. Within each safety design element, entities are encouraged to consider and document their use of industry standards, best practices, company

---

* https://www.eetimes.com/document.asp?doc_id=1333446&_mc=RSS_EET_EDT&utm_source=newsletter&utm_campaign=link&utm_medium=EETimesWeekInReview-20180721.
† https://www.theverge.com/2018/7/20/17595968/waymo-self-driving-cars-8-million-miles-testing.
‡ https://www.nhtsa.gov/sites/nhtsa.dot.gov/files/documents/13069a-ads2.0_090617_v9a_tag.pdf.

policies, or other methods they have employed to provide for increased system safety in real-world conditions. The 12 safety design elements apply to both ADS original equipment and replacement equipment or updates (including software (SW) updates/upgrades) to ADSs. However the NHTSA guidance is not specific enough to help manufacturers design effective safety mechanisms to reduce risk.

I.2 Characterizing the Safety of Automated Vehicles

How different is the concept or notion of safety in self-driving vehicles when compared to that used in other industries such as aviation, process control, and automotive? While the fundamental concepts are the same, the safety of self-driving vehicles has specific attributes that are different or not present in the safety of other industries. In this section we briefly discuss these attributes. When compared to the safety of traditional industries such as avionics, process control, and automotive, there are specific attributes pertaining to the safety of self-driving vehicles that we discuss next [23].

I.2.1 Performance Degradation

Traditional safety is based on faults and failures of mostly hardware (HW) components, and this is referred to as the reliability approach to safety [9]. In contrast, accidents involving self-driving vehicles might happen even if no HW device fails, but rather a performance degradation of some of its functions or intended functionality occurs. Addressing safety issues for these situations is referred to as safety of the intended functionality (SOTIF), and it is a fairly new concept as applied to the safety of self-driving vehicles [10]. Thus, some failures are due to performance degradation of self-driving vehicle components, typically involving higher levels of processing or higher levels of automation, for example, service failures. This definition of failure goes beyond that which is defined in the standard ISO 26262; however, it is compatible with other safety frameworks such as [8], System Theoretic Process Analysis (STPA) [11, 12, 13, 14, 15], or other real-time distributed systems [16]. One example of this understanding of the concept of safety is the failure of a vehicle detection system where the perception system provides missed detections (i.e., false negatives) or spurious detections (i.e., false positives). This could happen because the processing of environmental data is highly complex and the object detection function is subject to errors and impairments, particularly in bad weather or in night conditions when the visibility is poor. Either one of these failures could be catastrophic and could result in an accident or harm. Another example is a radio detection and ranging (RADAR) system correctly detecting objects only when the objects are moving, thus missing static objects because of limits on its performance. Thus, failure occurs in a degraded performance scenario.

I.2.2 **Focus on Software**

It is well known that the amount of SW in a vehicle continues a rapidly increasing trend that started with the development of by-wire systems. Much of the functionality of a self-driving vehicle is implemented in SW, and thus it is important to view the perception system as a set of SW servers, each providing services to the rest of the system. Therefore, one can refer to these various functionalities as a vehicle detection server, a pedestrian detection server, a road detection server, etc. The SW in self-driving vehicles is much larger in size and scope compared to traditional industries; thus, there should be a focus on the safety of the SW. As noted, a failure can occur if the SW services deviate from the correct services, and this could lead to safety hazards and safety risks. Ultimately, the overall safety of a self-driving vehicle will be dictated by the safety of its SW [9, 12].

I.2.3 **Non-Deterministic Perception System**

In the absence of HW faults, the perception systems of traditional industries are mostly deterministic in nature. For example, sensing the intake manifold pressure or engine speed in automotive systems is deterministic[4]. In contrast, the perception systems of self-driving vehicles are non-deterministic, leading to a high level of false positives and false negatives when their performance deteriorates to the point that service failures cannot be avoided. The non-deterministic aspect of the perception system stems from the fact that one never knows when its performance will deteriorate to the point where failures begin to appear in the services delivered by the system. Thus, the services provided by the perception system are subject to random failures, for example, when the weather deteriorates or when the system makes detection errors.

I.2.4 **Perception System Complexity**

Sensing elementary physical phenomena such as temperature or pressure is relatively simple, involving just some deterministic sensors, some electronics, and communications. In contrast, sensing or detecting man-made entities or constructs such as another vehicle, a road boundary, a city street, or a street intersection is complex because of the lack of structure of what is being sensed or perceived. The implication of the complexity of the perception system is that it is prone to errors, which degrades the performance or safety of the overall vehicle.

I.2.5 **Overall System Complexity**

In addition to its perception system, a self-driving vehicle also includes localization and mapping, planning and control, and actuation resulting in a highly complex system. One of the main issues with system complexity is that it makes testing for safety challenging, particularly if machine learning (ML) techniques are used, as it makes the design opaque to humans. This makes tracing the design and the test plans to the requirements problematic, since there is no human-understandable design that can be used for verification and testing [17]. In addition, it is known that when the system is complex, the system safety is affected by interacting complexity and tight coupling [9]. Another aspect of system complexity is that the autonomous vehicle operates in a complex external environment, and there are safety hazards due to events outside the domain of the autonomous vehicle, for example, from other vehicles (whether self-driving or not). Thus, the safety attributes of a self-driving vehicle are significantly different from those in other industries such as avionics, process control, and automotive.

In addition to its attributes, what are the various types of safety that encompass the overall safety of self-driving vehicles? As noted, the safety of self-driving vehicles is complex and differs from that of other industries such as avionics, process control, and automotive. On the one hand, there are safety commonalities such as the safety that involves component failures, which is the subject of so-called functional safety, and the safety involving components whose failure rates are well understood because they are proven in use, that is, in actual operation. On the other hand, there are two types of safety that are not prevalent in the avionics, process control, and automotive industries, and these include *SOTIF* and *multi-agent safety*. Thus, the types of safety that characterize the safety of self-driving vehicles include (1) traditional functional safety as defined by ISO 26262, (2) SOTIF, and (3) multi-agent safety [23]. Feth et al. also emphasize that safety assurance is a concern because established safety engineering standards and methodologies are currently not sufficient [22]. They also conclude that there are three types of safety that characterize the safety of self-driving vehicles: (1) traditional functional safety, (2) SOTIF which they assume are due to *functional insufficiencies*, and (3) multi-agent safety related to safe driving behaviors which are abstracted from technological challenges of situation awareness. Furthermore, they elaborate the fundamental safety engineering steps that are necessary to create safe vehicle of higher automation levels while mapping these steps to the guidance presently available in existing (e.g., ISO 26262) and upcoming (e.g., ISO PAS 21448 [26]) standards. Functional safety is a well-understood area which is guided by a number of international standards such as IEC 61508 [18], IEC 61511 [19], and ISO 26262 [6], and there are a large number of papers and publications on this topic. However, it is noted that ISO 26262 does not cover AVs; thus, its application should be done with great care. In the following, we characterize the safety category of SOTIF.

1.3 **SOTIF**

As noted, SOTIF is an important component of AV safety with incipient research, particularly at Society of Automotive Engineers (SAE) venues. As also noted, SOTIF addresses the safety aspects when dealing with hazards which are due to performance limitations and not necessarily due to faults. Some researchers [21, 22] consider performance limitations as *functional insufficiencies* defined as a deviation from intended functionality, that is, *behavior that vary from the original intent even if free of faults* [21]. There is an ISO standard working draft, ISO/WD PAS 21448 [26], that is targeted to SOTIF. From a top-level perspective, there are several major approaches to reduce risk due to hazards and due to performance limitations or functional inefficiencies. One approach is to use rigorous systems engineering precepts to ensure the safety of AVs and to provide a strong safety assurance case. Another approach is to incorporate technical solutions, techniques, and methods (e.g., fault-tolerant mechanisms) from appropriate disciplines to achieve the same goal. In this section we will summarize a paper that uses the first approach to ensure the safety of a pedestrian detection function using some principles of systems engineering. In Section 1.5, we will present one example of using the second approach, that of using technical safety mechanisms or measures to ensure safety.

Systems engineering uses the so-called V model to specify and describe key tasks while designing a complex or large system. These key tasks include requirements, design, implementation, testing, verification, and validation that are performed at specific times depicted as a *V* in Figure 1. The lifetime phases (or tasks) of requirements, design, and implementation are performed on the left side of the V, while the tasks involving testing, verification, and validation are performed on the right side of the V. The reason for depicting these activities on a V (rather than in a linear fashion) is that while working on the requirements specification, the corresponding specifications for validation are developed concurrently. While the completion of the requirements specification task is

FIGURE 1 Systems engineering V model listing main tasks on the left and right sides of the V.

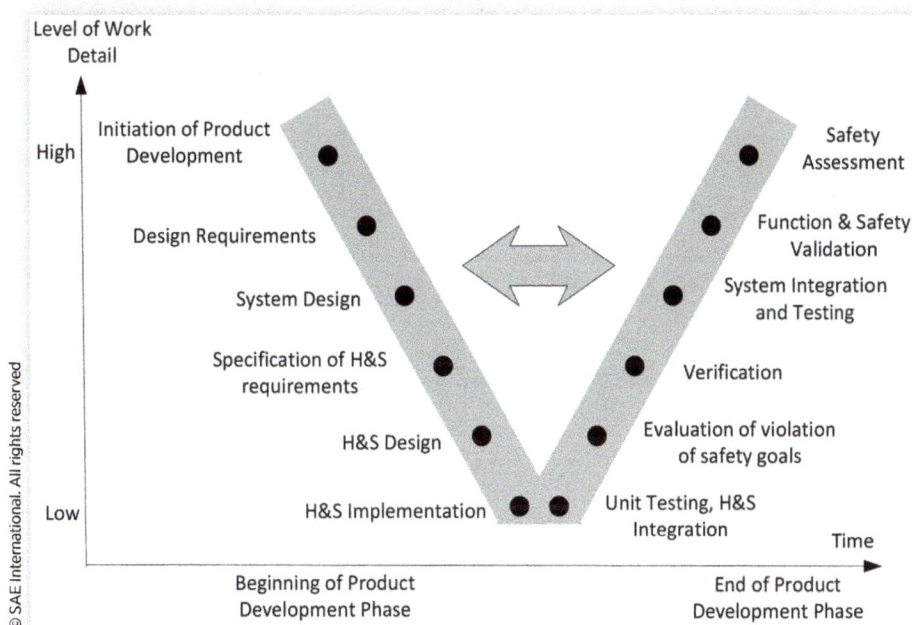

made very early in the development lifecycle of the product or project, the completion of the validation task is the last one. Gauerhof et al. [21, 27] describe how to structure validation targets to ensure safety of a pedestrian detection function in automated driving to address hazards due to performance limitations or functional insufficiencies. In this section we summarize their research.

The overall goal of Gauerhof et al.'s paper is to provide a strong safety assurance case for a *pedestrian detection function* implemented using ML techniques of artificial intelligence (AI). The ML technique described in the paper is based on CNNs (convolutional neural networks) that learn how to detect pedestrians using a large data set of examples. Providing a safety assurance case for an ML-implemented pedestrian detection function is challenging not only because we are dealing with functional insufficiencies but also because we are dealing with ML techniques which have serious gaps when used in safety-critical applications. The authors argue that to prevent functional insufficiencies (i.e., to address performance limitations) the set of *specifications must reflect the intended functionality* and that such specifications must be appropriate for any environment of vehicle operation. They also argue that what is needed is a set of well-structured and validated specifications of the intended functionality in all potential physical environments. According to the authors, there are two main sources for functional insufficiencies: (i) inherent uncertainty and complexity of the environment and (ii) intrinsic uncertainty within the functional implementation, for example, ML. In order to complete a thorough analysis of intrinsic uncertainties, the intended functionality has to be well understood and specified. The authors address three SOTIF hazard sources: under-specification, semantic gap, and deductive gap [21]. *Under-specification* occurs if the intended functionality is more diverse than what is specified. Addressing under-specification by means of generalization can lead to an inadequately defined set of safety requirements. *Semantic gap* refers to using implicit knowledge on the satisfaction of safety goals. *Deductive gap* involves using invalid assumptions on different abstraction levels causing unintended functionality. The authors further argue that to correctly specify the functions with intrinsic uncertainty, one must have expert knowledge about failure conditions or the ground truth. Unfortunately, for some situations, for example, SAE automation levels 4 and 5, it is extremely challenging and difficult to have ground truth.

The main contribution of Gauerhof et al.'s paper is the development of a set of well-structured validation targets to demonstrate that an ML-implemented pedestrian detection function fulfills its intended functionality. The pedestrian detection function consists of two sub-tasks: classification and localization. For each pedestrian class, they propose the following functional requirements that need to be validated at a later stage:

- *Pedestrian of height (X1 pixels) and of width (X2 pixels) are classified.*
- *Pedestrians are detected if Y % of the person is concealed.*

- *There are less than W1 false positives per 1000 frames.*

- *There are less than W2 false negatives per 1000 frames.*

- *There are less than B1 misclassified detections.*

- *Confidence level shall reflect the actual uncertainty of correctness of a classification.*

In terms of reducing risk due to hazards caused by under-specification, the authors suggest the following validation targets:

- *Environment is sufficiently well known.*

- *Task is sufficiently well known. The following evidence is provided: Requirements shall be specified including task-specific attributes. In the case of ML generalization abilities, attributes such as color invariances and translation invariances might be required.*

- *Sensitivity against unpredictable or unspecified impact of environmental attributes is sufficiently low. The following evidence is provided: Sensitivity to environmental changes shall be investigated. Moreover, influence due to distributional shift over time or due to geographic changes shall be reviewed. Requirements on invariance and generalization attributes shall be reviewed according to their appropriateness to the intended functionality. Run-time monitoring of assumptions and field-based validation shall be used to investigate discrepancies between the real environment and the assumptions as well as sensitivity to these changes. Moreover, statistical extrapolation shall be used to generalize the results of acquired data.*

In the context of ML, one issue while reducing risk due to hazards caused by semantic gaps is making claims on the relevance of references used for training, tests, and validation of data sets. To address these issues, the authors suggest the following sub-goals:

- *Pedestrian classes are sufficiently accurately described. Evidence: Functional specification of several validation data subsets shall include all variants of classes that can be derived from the environment. Moreover, safety requirements shall be transferred into task-specific requirements, for example, informal textual specifications shall be transferred into formal specifications as far as it is possible, at least for safety-relevant requirements. Evaluation of specific influences and appropriate object variations shall be specified beyond statistical evaluation.*

- *Location accuracy is sufficiently well described. Evidence: Training and validation data shall be specified. Evaluation of specific influences shall be specified. Additionally, evaluation of compliance with tolerances, of size, and of location variation shall be specified.*

- *Discrepancy between real and described environment is sufficiently small. Evidence: Evaluation of similarity between reality and specification of validation data shall be specified. Functional modifications, such as run-time monitoring, degradation modes, preprocessing of ML input, etc., shall be specified and documented.*

In the context of ML, the following issues are identified while reducing risk due to hazards caused by deductive gaps: (i) classification features might be wrongly learned and (ii) feature classifiers might be erroneously implemented. To address these issues, the authors suggest the following validation targets to be used before and during training of the neural network:

- *Data set is sufficient for the intended functionality. Evidence: Transfer of system-level requirements to ML-specific requirements as well as the attribute distribution within training, test, and validation data sets shall be evaluated. Moreover, independence from unintended object relations shall be highlighted. For example, synthesized data can be used to broaden recorded data by special attributes.*

- *Overfitting is sufficiently reduced. Evidence: Overfitting measures, such as pretraining on diverse data set, regularization methods, Dropout or DropConnect, and data augmentation, shall be documented and evaluated.*

- *Underfitting is sufficiently reduced. Evidence: Underfitting measures (e.g., a minimum amount of training data for each class variant) shall be documented and evaluated.*

To evaluate weaknesses of ML-implemented functions, the following methods are well known: feature visualization, structuring of the input space, formal verification, and dealing with uncertainty. In terms of the latter, the confidence level of each class output does not express a probability of existence of the object itself. Therefore, uncertainty calculation might be used to measure the reliability of the classification result. Uncertainty quantification can be used for further measures (e.g., in plausibility checks and sensor fusion algorithms), thus improving the overall robustness and reliability of the subsequent trajectory planning tasks. Based on the above, the paper suggests the following evidences:

- *Essential influences on the ML function are sufficiently understood. Evidence: The application of feature visualization, adaptation of confidence level, and uncertainty calculation shall be documented. Furthermore, correlations of errors to features shall be investigated and reduced by appropriate training. Evaluation of these correlations shall be documented.*

- *ML function is sufficiently robust. Evidence: Tolerance against distributional shift and adversarial and faulty input shall be evaluated. Statistical evaluation shall be documented. An integrity test of ML function shall be documented.*

- *Learned features are sufficient for function. Evidence: Learned features and correlations between these and detection results shall be analyzed and documented (e.g., by feature visualization).*

Taking into account well-known weaknesses of ML such as *sensitivity to adversarial attacks*, the authors suggest the following set of validating targets during system development:

- *Changes to parameters do not inviolate safety requirements. Evidence: Verification specification for any changes shall be documented.*

- *Differences between the training and target platforms do not lead to a violation of the safety requirements. Evidence: Verification specification for any changes shall be documented.*

- *Changes in target platform comply with safety requirements. Evidence: Verification specification for any changes in target platform shall be documented.*

In summary Gauerhof et al.'s paper provides a set of validation targets designed to provide a strong safety case to address safety hazards caused by functional insufficiencies (i.e., performance limitations), particularly those arising in the perception system of an AV.

I.4 Integrated Approach to the Safety of Automated Vehicles

As postulated in this series of edited books by SAE International, there are three main perspectives of safety when discussing the safety of AVs: functional safety, SOTIF, and multi-agent safety. At the highest level, the underlying concepts and understanding of safety are the same, *the absence or reduction of risk which can lead to accidents or harm to people*. However, at lower levels of abstraction, the approach of these safety perspectives when applied to autonomous vehicle safety is not the same, particularly when discussing multi-user safety. There are a large number of papers about the safety of AVs taking into account other agents in a dynamic fashion (i.e., other vehicles, pedestrians, moving obstacles, etc.). However, most of the literature address multiple-agent safety in an ad hoc manner and not integrated with the other safety categories of functional safety and SOTIF. Can all of these approaches or views be reconciled? Is there a unique framework for discussing the safety of AVs? In this section, we propose such a framework for integrating and unifying the discussion of safety at lower levels of abstraction when viewed from these three safety perspectives.

The unifying framework is based on dependability theory that includes fundamental principles of not only functional safety but reliability, availability, and security and which are based on cause and effects, that is, causal relationships [8]. According to functional safety precepts, the fundamental causes or sources of accidents are faults, errors, and failures which can lead to hazards, pose risks, and can potentially lead to accidents. This chain of events is depicted in Figure 2. When untreated, these sources may represent a system that is highly unsafe. On the other hand, when properly treated and dealt with, the root causes of accidents can be treated or managed appropriately representing a highly safe system. We adopt the definitions of safety concepts drafted by well-known safety standards such as ISO 26262 [6]. We define *fault* as an "abnormal condition that can cause an element or an item to fail and it can be classified as permanent or intermittent." An *error* is a "discrepancy between a computed, observed or measured value or condition, and the true, specified or theoretically correct values." An error can arise as a result of

FIGURE 2   Chain of safety events leading to an accident.

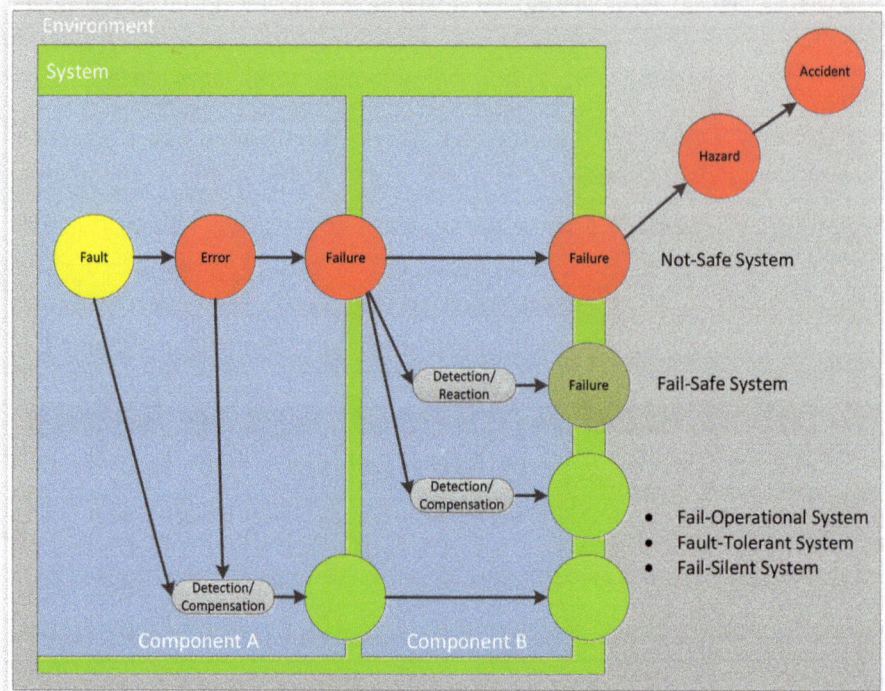

unforeseen operating conditions or due to a fault. A *failure* is the "termination of the ability of an element or item to perform a function as required."

A *safety hazard* is defined as the "potential source of harm or accident caused by malfunctioning behavior of an item." A hazard is an abstract concept that needs the element to be protected in order to be realized or materialized and causes harm or accidents. When people or resources are exposed to hazards, the ensuing situation is called *risk*. There are many levels of risk, and to properly distinguish them, we need to know the *probability of exposure* to the hazard, the *severity* of the ensuing harm or accident, and the degree of avoidance of the specific harm or damage through the timely reactions of the drivers involved (called *controllability*). The following equation conveys this notion of various levels of risk:

$$R = S \times PE \times PC \tag{1}$$

where R is the safety risk, PE is the probability of hazard occurrence (i.e., exposure), S is the hazard severity, and PC is the probability of controllability of the situation.

Functional safety is defined as the "absence of unreasonable risk due to hazards caused by malfunctioning behavior of E/E systems." Thus functional safety deals only with malfunctions (mostly faults) of the very equipment that is brought in to enable automation, that is, electrical/electronic (E/E) systems. From this definition, it is clear the functional safety does not deal with hazards caused by temporal performance limitations (in the case of SOTIF) or hazards caused by other drivers (whether human or automated) or other agents as is the case in multi-user safety. As noted, when undetected or untreated, faults, errors, and failures lead to an unsafe system, as illustrated in the top portion of Figure 2. However, when detected, compensated, and treated in a reactive or proactive fashion, it is possible to design systems with various levels of safety ranging from a fail-safe system, to a fail-operational system, to a fault-tolerant system, to a fail-silent system as depicted in the lower portions of Figure 2.

I.4.1 Risk-Based Approach to Multi-Agent Safety

There exists a great deal of literature on multi-agent safety for AVs at the various SAE levels of automation. However, most of this literature discusses safety in an ad hoc manner that is not consistent across the industry and also not consistent with the other safety perspectives, particularly functional safety which follow a risk-based approach. In the following we discuss a model useful to discuss multi-user safety using a risk-based approach and thus in a way that can be integrated with the other safety perspectives, that is, functional safety and SOTIF. The model, shown in Figure 3, depicts a unifying approach for modeling the root causes of accidents of the three main AV safety categories: functional safety, SOTIF, and multi-user safety. We do not detail discussions on functional safety as this has been discussed extensively in the literature,

FIGURE 3 Unifying model for integrating SOTIF, multi-user safety, and functional safety.

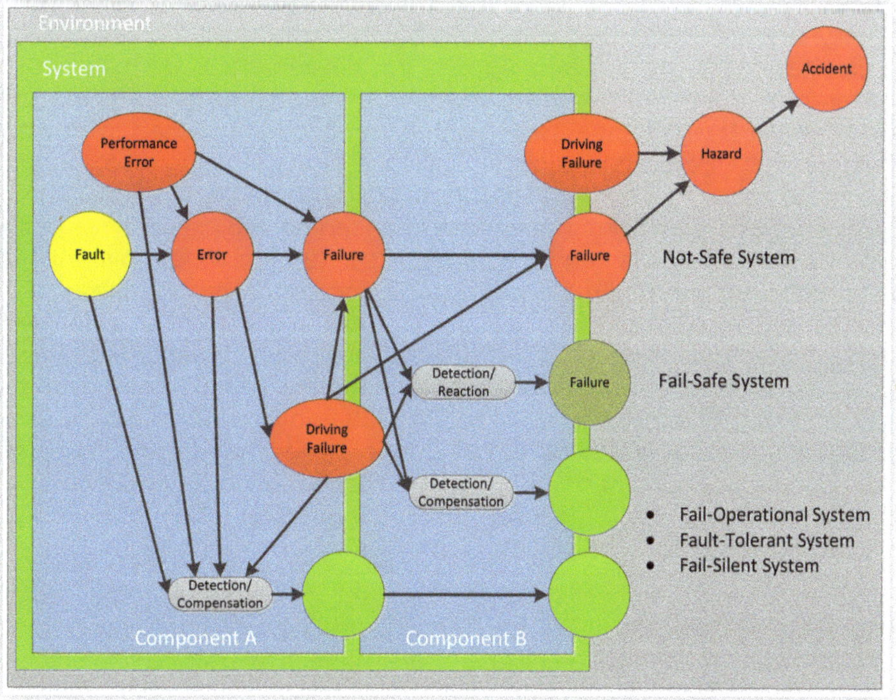

rather we concentrate on discussing about SOTIF and multi-user safety. The model is basically an extension of the model of Fig. 2 that depicts the chain of events leading to accidents from the viewpoint of functional safety.

In terms of *SOTIF*, performance limitations of some functional units (e.g., the perception system of AVs) can lead to errors which in turn can lead to failures thus constituting hazards and eventually leading to accidents. A good example is performance limitations of a light detection and ranging (LIDAR) sensor in bad weather, particularly when dealing with rain and fog, leading to false positives and false negatives. Thus, even perfectly functioning (i.e., no HW and SW faults) LIDAR units can be the source of errors which could lead to accidents. With this understanding of error and with additional definitions of some concepts, the entire theory and application of safety developed for functional safety can be used for modeling and discussing SOTIF. To reconcile SOTIF within the framework of functional safety, one can add another definition of *error* as the "discrepancy between a computed, observed or measured value or condition, and the true, specified or theoretically correct values strictly due to performance limitations of the measuring apparatus." Performance limitations are modeled in Figure 3 as *performance errors* which can lead to other errors and failures that can potentially cause accidents.

In terms *of multi-user safety*, the actions of other agents in the road or the actions of the computing system driver (whether a human or auto-mated or whether a backup driver is involved) of the AV can be the source of failures or hazards that can lead to accidents as shown in Figure 3. This framework needs additional definitions and clarifications of the concepts of fault, error, failure, hazard, and risk so that they are consistent with those in functional safety. For the case of multi-user safety, another definition of *driving failure* can be added as follows: "termination of the ability of driver or the vehicle system to avoid hazardous situations." Likewise, another definition for *safety hazard* can be added as the "potential source of harm or accident caused by driving failures." As an example, consider the case of maintaining a *safe longi-tudinal distance* to a leading vehicle to avoid accidents. It is well known that there exists a minimum distance to a leading vehicle so that a colli-sion with such vehicle can be avoided [20]. Such minimum distance is a function of the maximal braking and acceleration commands, the response time of the driver, the longitudinal velocities of the involved vehicles, and their lengths [20]. If such safe minimum distance is violated by either a human or automated driver, then the vehicle is exposed to risk that might lead to an accident. A violation on maintaining a safe distance to a leading vehicle can be due to an error in the calculation of safe distance (an error condition), driver distraction (a driving failure), a bug in the SW (a driving failure), etc. The associated chain of events is shown in Figure 3.

Thus, by adopting a more general model for the underlying sources of accidents as depicted in Figure 3 and by adding additional definitions and concepts related to faults, errors, and failures, we can have a unified view of the safety of AVs that covers the three categories: functional safety, SOTIF, and multi-user safety. In summary, for the case of functional safety, HW or SW faults lead to errors and failures, resulting in safety hazards. For the case of SOTIF, performance degradation in the perception system can cause errors leading to failures involving safety hazards. Finally, for the case of multi-user safety, human or computing system-based drivers can cause failures leading to safety hazards.

I.4.2 HARA and Risk Reduction

The ISO 26262 standard details the phases of HARA (hazard analysis and risk assessment) and risk reduction which are critical to designing safe systems from the viewpoint of functional safety. Addressing HARA and risk reduction to designing safe systems from the viewpoint of SOTIF and multi-agent safety is not straightforward. This is so because in the case of functional safety this is done using Automotive Safety Integrity Level (ASIL) classifications for the various system components and designing specific safety mechanisms (i.e., risk reduction) according to the ASIL levels, and in the case of SOTIF and multi-user safety, the industry has not yet defined corresponding ASIL levels.

In the case of functional safety, ASIL is determined as a function of probability of exposure to the hazard, the severity of the potential harm or accident, and the controllability, as can be seen from Equation (1). This is done using 3 levels for severity, 4 levels for exposure, and 3 levels for controllability resulting in 3x4x3 = 64 levels for ASILs and classified as QM, ASIL A, ASIL B, ASIL C, and ASIL D with the last one being the most safety-critical. For traditional automotive applications, we have a wealth of knowledge and techniques on how to determine or estimate levels for severity, exposure, and controllability and thus how to estimate ASIL levels. No such set of knowledge and techniques exist for SOTIF and multi-agent safety, and we need much more research and work in this area in order to reconcile them with the wealth of knowledge that exists for functional safety.

I.5 SOTIF Safety Measures and Mechanisms

The use of appropriate standards (e.g., ISO 26262) or systems engineering concepts such as validation target cannot guarantee that hazards will not occur or that risk reduction efforts will be effective. We need to combine the above with technical measures or mechanisms that proactively deal with the hazards and help reduce risk to acceptable levels. In the context of the pedestrian detection function, analyzed by Gauerhof et al. [21, 27], the authors suggest some measures at the functional and system levels that will help reduce risk induced by functional insufficiencies. The suggested safety measures at the functional level include preprocessing of the ML input according to known factors that significantly influence performance and post-processing of the ML output to include adjustments of confidence levels based on factors known to influence performance, so that decisions about driving behavior and trajectory planning are adapted to the reliability of the perception function. The suggested safety measures at the system level include [21]:

- *Diverse redundancy increases the dependability of a function. For pedestrian detection, several possibilities exist, for example, LIDAR, RADAR, and traditional computer vision algorithms.*

- *Operating modes, also called degradation modes, depend on the vehicle's environmental model. As long as the environmental model is reliable, decisions are taken within a wider range of possible trajectories. In contrast a degradation mode is chosen according to a cautious and defensive driving strategy, if an object is detected with a low confidence level.*

- *Transition between operating modes ensures a continuous driving behavior.*

- *Run-time monitoring of assumptions allows the validation of whether assumed attributes about environment are still valid. The detection of discrepancies between distribution of environmental attributes and design assumptions at run-time could indicate either errors in the trained function or that the system is operating within an environment for which it was not adequately trained.*

- *Established driver assistance systems (e.g., emergency brake assist) applied to AD must be reviewed from a systems engineering perspective. It must be clarified to what extent measures must be taken at the system level to reduce the integrity requirements on the individual functional components.*

Other examples of safety measures that will help in reducing risk caused by functional insufficiencies related to a vehicle detection function involve fault-tolerant techniques and fault-containment protocols such as the one discussed by Pimentel and Bastian [23] and Pimentel, Bastian, and Zadeh [24]. In the following we summarize the safety measures and mechanisms proposed in [23].

Once safety hazards and risks are identified and analyzed to improve safety, techniques must be used to reduce risk to an acceptable level. Reducing safety risks of self-driving vehicles is achieved by using fault-tolerant design methodologies that include component replication and safety mechanisms and safety measures [16]. Fault tolerance is the ability to deliver a specified functionality in the presence of one or more specified faults. In this section we discuss the ensuing terms and propose a fault-tolerant architecture that makes extensive use of replicated components. Redundancy is the duplication of characteristics of a critical element with the intention of performing a required function or to represent information. Redundancy is used to achieve a safety goal or a specified safety requirement or to represent safety-related information. Duplicated functional components can be an instance of redundancy for the purpose of increasing availability or allowing fault detection. For example, the addition of parity bits to data representing safety-related information provides redundancy for the purpose of allowing fault detection. A safety mechanism is a technical solution implemented by electrical and/or electronic functions or elements, or by other technologies, to detect and mitigate or tolerate faults, or control or avoid failures in order to maintain intended functionality or to achieve or maintain a safe state.

Safety mechanisms are implemented within an item to prevent a fault from leading to single-point failures or to reduce residual failures and prevent faults from being latent. The safety mechanism is either (a) able to transition to, or maintain, the system in a safe state or (b) able to alert the driver such that the driver is expected to control the effect of the failure, as defined in the functional safety concept. One or several safety mechanisms make up a safety measure, which is defined as an activity or technical solution to avoid or control systematic failures, and to detect or control random HW failures, or mitigate their harmful effects. Unreasonable risk is risk that is judged to be unacceptable in a certain societal context.

I.5.1 Fault-Tolerant Perception System

Typical perception system sensors and actuators for AVs provide their outputs (or inputs) via Ethernet, controller area network (CAN), CAN flexible data rate (CAN-FD), and universal serial bus (USB) communication drivers that enable interfacing to all sensors of the perception system. These sensors include cameras, LIDAR units, RADAR systems, ultrasonic sensors, inertial measurement units (IMUs), etc. The communication drivers for Ethernet, CAN, and USB typically interconnect with computing units such as the central processing unit (CPU), graphics processing unit (GPU), and field-programmable gate array (FPGA) through a peripheral component interconnect express (PCIe) bus as shown in Figure 4. It is well known that the safety of a system can be improved by using replicated components, and in particular

FIGURE 4 Computing, communications, sensors, and actuator components of a perception system. S, sensor; A, actuator; CAN, controller area network; FPGA, field-programmable gate array; GPU, graphics processing unit.

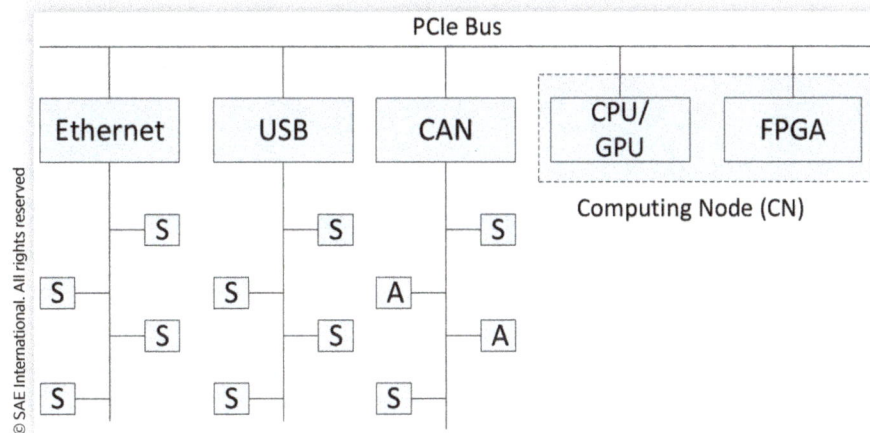

Automated Vehicles: Safety of the Intended Functionality (SOTIF)

FIGURE 5 Replicated architectural components: physical depiction. E, Ethernet; U, universal serial bus; C, controller area network; CN, computing node.

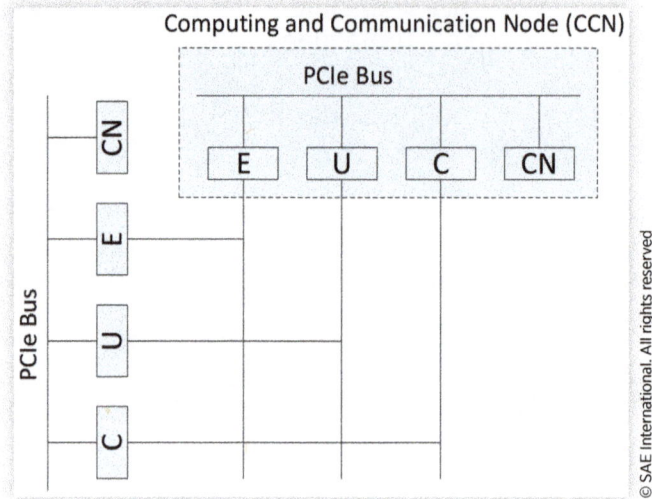

Computing and Communication Node (CCN)

PCIe Bus

CN

E

U

C

PCIe Bus

E U C CN

FIGURE 6 Replicated architectural components: logical depiction. E, Ethernet; U, universal serial bus; C, controller area network; CCN, computing and communication node.

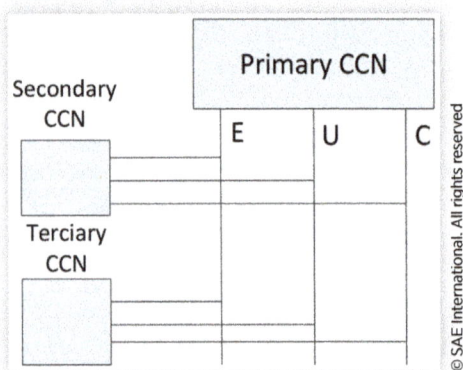

Primary CCN

Secondary CCN

E U C

Terciary CCN

using active replication, where all servers within a group execute the same service requests in parallel and execute an agreement protocol to agree on the outputs given the same inputs, assuming no-fault conditions [16].

For perception systems of self-driving vehicles, the following can be replicated: computing nodes, communication and computing nodes, communication lines, sensors, and actuators. Although any component can be duplicated or triplicated as shown in Figures 5 and 6, only the most safety-critical components or modules should have redundant or replicated components, and these include computing elements, the perception system, and communication features. Intuitively, the system safety is improved when active replication is used, because whenever there is a failure of any element, its replica is enabled to provide the same services as that which failed, and thus the system continues normal operation. Obviously, replicating components, particularly expensive sensors such as LIDARs, increases the cost, and thus a careful balance must be made to trade safety for cost.

1.6 **Fault-Containment Protocol**

As noted, when the system is complex, the system safety is affected by interacting complexity and tight coupling, and this leads to error and failure propagation. One way to contain the propagation of errors and failures, and thus to reduce risk, is to use a fault-containment protocol.

In this section we propose a fault-containment protocol suitable for applications involving vehicle detection. Fault containment is one of the main methods of risk reduction. The idea is to detect the fault and minimize its effects on the remainder of the system and possibly isolate the failed component, service, or module. The fault-containment protocol has three main components, and it is organized hierarchically: the perception system model, the controller model, and error detection. The fault-containment protocol runs at the top level (controller model) and makes use of the perception system model and error detection activity to model perception system failures.

I.6.1 Perception System Model

The perception system of an autonomous vehicle performs many functions, such as detecting other vehicles, lanes, roads, and pedestrians, along with many other functions. These functions are not performed perfectly, and thus they stand to benefit from safety measures in the SOTIF category. To provide failure-tolerant designs, we model the behavior of any safety-critical sensor in the perception system as depicted in the state transition diagram of Figure 7. The perception system model starts in a normal operation state where there are no errors. When there is an error in the vehicle detection function, the system goes to the transient failure state, where the error is monitored for a time interval τ. If no further errors are reported within the interval τ, then the system returns to the normal operation state; otherwise the system enters the semipermanent failure state and a vehicle detection error counter (ECvdet) is incremented. From the semipermanent failure state, the system returns to the normal operating state if the vehicle detection function is reset or restarted, for example, by switching the vehicle off and restarting it again. However, if the vehicle detection error counter exceeds a limit value, the perception system is considered to have a permanent failure.

FIGURE 7   State transition model of a safety-critical perception system component. EC: error counter.

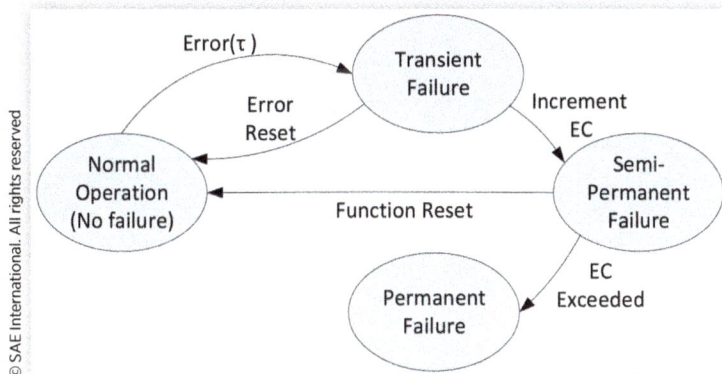

I.6.2 **Controller Model**

The associated fault-containment model within the vehicle control system dealing with vehicle detection is shown in Figure 8. The safety controller state, assuming error-free operation, switches between the states' normal operation and vehicle detected. While in the vehicle-detected state, the controller performs a safe control action and returns to the normal operation state. When an error occurs, either a missed vehicle detection or a spurious vehicle detection, the controller goes to the corresponding state. Upon entering the missed detection or spurious detection states, the controller performs an emergency control action to bring the controller state to a fail-safe state in an effort to avoid an accident or reduce harm. While in the vehicle-detected state, the controller might perform an unsafe control action (UCA) that leads to the fail-safe state. From the latter state, the controller actions may result in an accident, depending upon the overall vehicle situation and environmental conditions at the time the fail state is entered.

I.6.3 **Error Detection**

As can be seen from Figures 7 and 8, the fault-containment protocol requires that errors are detected. Detecting perception system HW sensor faults is much easier than detecting SW faults or errors; for example, correctly recognizing or identifying spurious detections or missed detections is extremely difficult. Detecting SW errors requires an advanced understanding of the algorithms used in the underlying SW [25]. For the case of vehicle detection, if the detection SW is based on ML algorithms such as deep learning (DL), then, to the authors' knowledge, this is an unsolved problem, as we as a research community do not completely understand how DL algorithms actually perform object detection due to the "end-to-end" nature of the process [30]. For this

FIGURE 8 State transition diagram of a safety-critical control system component. SCA, safe control action; UCA, unsafe control action; ECA, emergency control action.

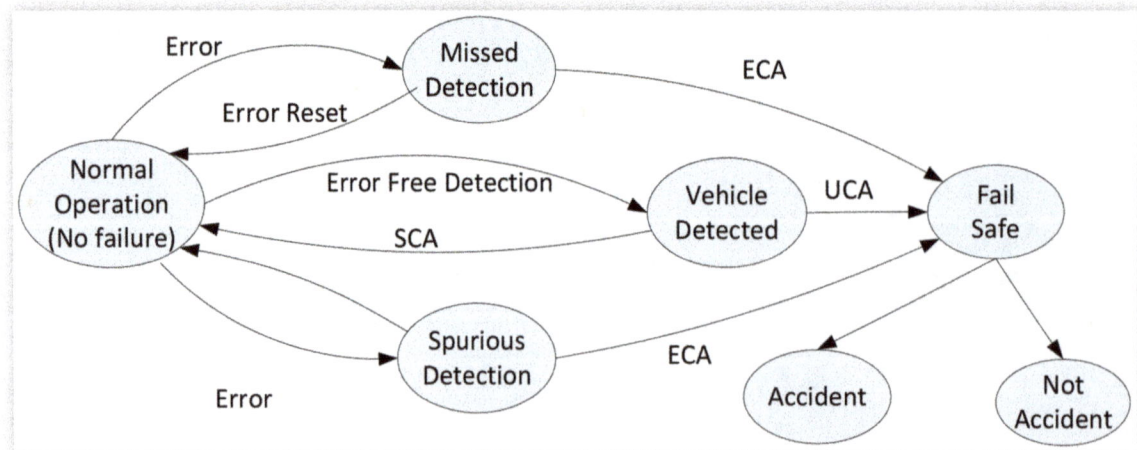

reason, neural network models are considered nontransparent [28]. If the detection SW is based on model-based computer vision algorithms [29], for example, involving histograms of oriented gradients (HOG) feature extraction, the training of linear support vector machine (SVM) classifiers, and using the sliding-window techniques for estimating a bounding box for the detected vehicles as used in [30], then a form of N-version programming can be used for error detection [16].

One way to detect perception system errors is using a set of key performance indicators (KPIs) to keep track of the confidence level of the vehicle detection system as to its final conclusions, that is, vehicle detected or no vehicle detected. The KPIs can be a function of some of the parameters of the algorithm stages, for example, HOG feature extraction, SVM classifier sliding window, bounding boxes, etc. If the confidence level is low, then an error can be assumed. The error rate can be decreased considerably if redundancy is used; more specifically, if several replicas are employed, then a majority voter can be used. Regardless of the soft computing approaches used and the evaluation of their outputs, however, detecting errors and faults in autonomous vehicle perception systems is challenging and research is just beginning [31].

1.7 **The Papers in This Collection**

There are not a large number of papers in the area of SOTIF from SAE and other sources. The following is a representative list of ten SAE papers on the subject.

1. Jinpeng Xu and Feng Luo, "Fault-Tolerant Ability Testing for Automotive Ethernet," SAE Technical Paper 2018-01-0755, 2018, doi:10.4271/2018-01-0755.

2. Rick Salay, Rodrigo Queiroz, and Krzysztof Czarnecki, "An Analysis of ISO 26262: Machine Learning and Safety in Automotive Software," SAE Technical Paper 2018-01-1075, 2018, doi:10.4271/2018-01-1075.

3. Junfeng Yang, Michael Ward, and Jahangir Akhtar, "The Development of Safety Cases for an Autonomous Vehicle: A Comparative Study on Different Methods," SAE Technical Paper 2017-01-2010, 2017, doi:10.4271/2017-01-2010.

4. William Buller, Helen Kourous, and Jakob Hoellerbauer, "Autonomous Vehicle Sensor Suite Data with Ground Truth Trajectories for Algorithm Development and Evaluation," SAE Technical Paper 2018-01-0042, 2018, doi:10.4271/2018-01-0042.

5. Dajiang Suo, Sarra Yako, Mathew Boesch, and Kyle Post, "Integrating STPA into ISO 26262 Process for Requirement Development," SAE Technical Paper 2017-01-0058, 2017, doi:10.4271/2017-01-0058.

6. Oleg Lurie and Joseph Miller, "Hazard Analysis and Risk Assessment beyond ISO 26262: Management of Complexity via Restructuring of Risk-Generating Process," SAE Technical Paper 2018-01-1067, 2018, doi:10.4271/2018-01-1067.

7. Philip Koopman and Michael Wagner, "Toward a Framework for Highly Automated Vehicle Safety Validation," SAE Technical Paper 2018-01-1071, 2018, doi:10.4271/2018-01-1071.

8. Mario Berk, Hans-Martin Kroll, Olaf Schubert, Boris Buschardt, and Daniel Straub, "Bayesian Test Design for Reliability Assessments of Safety-Relevant Environment Sensors Considering Dependent Failures," SAE Technical Paper 2017-01-0050, 2017, doi:10.4271/2017-01-0050.

9. Philip Koopman and Michael Wagner, "Challenges in Autonomous Vehicle Testing and Validation," *SAE Int. J. Trans. Safety* 4(1):2016, doi:10.4271/2016-01-0128.

10. Philip Daian, Bhargava Manja, Grigore Rosu, Shinichi Shiraishi, and Akihito Iwai, "RV-ECU: Maximum Assurance In-Vehicle Safety Monitoring," SAE Technical Paper 2016-01-0126, 2016, doi:10.4271/2016-01-0126.

In the following, we summarize the above papers in the context of the characterization of the safety of self-driving vehicles. After this introduction, the actual papers follow.

I.7.1 Discussion of the SOTIF Papers

1. Jinpeng Xu and Feng Luo, "Fault-Tolerant Ability Testing for Automotive Ethernet," SAE Technical Paper 2018-01-0755, 2018, doi:10.4271/2018-01-0755.

There is a strong interest in the industry to use automotive Ethernet to supplement and even to replace the CAN bus protocols for some safety-critical applications. With the introduction of BroadR-Reach and time-sensitive networking (TSN), Ethernet has become an option for in-vehicle networks (IVNs). Accordingly, to make a strong safety case, a great deal of testing, verification, and validation must be made using systems engineering precepts. In this paper, the authors report on several tests to verify some fault-tolerant features of automotive Ethernet. In summary, their testing results showed (a) Ethernet requires better connections, (b) Ethernet has better resistance to ground shift, and (c) Ethernet wires should avoid short circuits with high-voltage pins. The authors further conclude that Ethernet has less tolerance for connector and wire aging, which means better components are required to guarantee their physical properties during the lifecycle of the system. With the help of ESD components, Ethernet shows better tolerance for faults when a wire shorted to BAT. Although some time is needed for recovery from the fault, a continuous short-circuit state does not result in long communication failures. The lost frames can be retransmitted by the application, or transmitted with a redundant link, which ensures that the failure has a minimal impact on communication. Other faults may lead to frame loss or stopped communication, but not permanent damage. Each result shows equal or better fault tolerance compared with CAN, which means automotive Ethernet has the potential to be more widely used with better connections and improved circuits. Although these test results are related to chip manufacturers and production batches, the analysis results still have guiding significance for the application of automotive Ethernet to avoid dangerous conditions.

2. Rick Salay, Rodrigo Queiroz, and Krzysztof Czarnecki, "An Analysis of ISO 26262: Machine Learning and Safety in Automotive Software," SAE Technical Paper 2018-01-1075, 2018, doi:10.4271/2018-01-1075.

Providing a strong safety assurance case for AV functions, for example, *pedestrian detection* implemented using ML techniques of AI, is challenging. This is so not only because we are dealing with functional insufficiencies but also because we are dealing with ML techniques which have serious gaps when used in safety-critical applications. The ISO 26262 standard for functional safety of road vehicles provides a comprehensive set of requirements for assuring safety but does not address the unique characteristics of ML-based SW. In this paper, the authors provide a set of suggestions in several areas in order to address this gap. In terms of identifying hazards, the authors recommend that ISO 26262 expands their definition of hazard to include those due to functional insufficiencies. In terms of fault and failure modes, they recommend that ISO 26262 be extended to explicitly address the ML lifecycle and require the use of fault detection tools and techniques that are customized to this lifecycle. This is because ML components have a development lifecycle that is different from other types of SW and analyzing the stages in the lifecycle reveals distinct types of faults they may have. In terms of specification and verification, they recommend that ISO 26262 provides different safety requirements depending on whether the functionality is specifiable. This is because ML components are trained from inherently incomplete data sets, and they violate the assumption in V model-based processes that component functionality must be fully specified and that refinements are verifiable. Furthermore, it is possible that certain types of advanced functionality (e.g., requiring perception) for which ML is well suited are unspecifiable in principle. As a result, ML components are designed with the knowledge that they have an error rate and that they will periodically fail. In terms of the level of ML usage, they recommend that ISO 26262 only allow the use of ML at the component level. This is because end-to-end approach challenges the assumption that a complex system is modeled as a stable hierarchical decomposition of components each with their own function and this limits the use of most techniques for system safety. In terms of required SW techniques, they recommend that the requirements be expressed in terms of the intent and maturity of the techniques rather than their specific details in order to remove an identified bias. ISO 26262 mandates the use of many specific techniques for various stages of the SW development lifecycle. Analysis shows that while some of these remain applicable to ML components and others could readily be adapted, many remain that are specifically biased toward the assumption that code is implemented using an imperative programming language.

3. Junfeng Yang, Michael Ward, and Jahangir Akhtar, "The Development of Safety Cases for an Autonomous Vehicle: A Comparative Study on Different Methods," SAE Technical Paper 2017-01-2010, 2017, doi:10.4271/2017-01-2010.

    To achieve complete safety of an AV, a safety case providing guidance on the identification and classification of hazardous events and the minimization of these risks needs to be developed throughout the entire development lifecycle process. A comprehensible and valid safety case has to employ appropriate safety approaches complying with the automotive functional safety requirements in ISO 26262. A valid safety case for an AV consists of four main interdependent components, namely, (a) safety target that must be addressed to assure vehicle safety; (b) evidence for the safety target obtained from study, analysis, and test of the vehicle system; (c) argument showing how the rationale indicates compliance with the safety target; and (c) context identifying the basis for the argument presented. Based on the goals of the safety work, the principles of safety process rationale argument are hazard generation, risk assessment, and hazard management, that is, addressing safety requirements through an appropriate combination of system design in accordance with the ASIL indicated. This paper presents a comparative study of different safety approaches, in particular, failure mode and effects analysis (FMEA) method and goal structuring notation (GSN) method, that have been employed to generate lists of hazardous events, safety goals, and functional safety requirements at the vehicle level. A case study on the safety case development of INSIGHT autonomous vehicle has been carried out using the aforementioned methods. This case study covers the safety argument of battery and charging system that supply the whole electric power for INSIGHT vehicle. The safety of this system has been assessed along with their potential for malfunction together with the layers of protection. The results and conclusions from case study analyses suggest the safety case of CAVs can be developed in a highly effective manner by employing a combined method of GSN and FMEA.

4. William Buller, Helen Kourous, and Jakob Hoellerbauer, "Autonomous Vehicle Sensor Suite Data with Ground Truth Trajectories for Algorithm Development and Evaluation," SAE Technical Paper 2018-01-0042, 2018, doi:10.4271/2018-01-0042.

    One of the principal bottlenecks for sensing and perception algorithm development is the ability to evaluate tracking algorithms against ground truth data. By ground truth it is meant an independent knowledge of the position, size, speed, heading, and class of objects of interest in complex operational environments. This paper describes a multi-sensor data set, suitable for testing algorithms to detect and track pedestrians and cyclists, with an autonomous vehicle's sensor suite. These data can be leveraged to evaluation algorithms ranging from sensing, perception, detection, tracking, prediction, and classification. The data set can be used to evaluate the benefit of fused sensing algorithms and provides ground truth trajectories of pedestrians, cyclists, and other vehicles for objective evaluation of track accuracy. Our goal was to execute a data collection campaign at an urban test track in which trajectories of moving objects of interest are measured with auxiliary instrumentation, in conjunction with several autonomous vehicles (AV) with a full sensor suite of RADAR, LIDAR, and cameras. Multiple autonomous vehicles collected measurements in a variety of scenarios designed to incorporate real-world interactions of vehicles with bicyclists and pedestrians. Trajectory data for a set of bicyclists and pedestrians was collected by separate means. In most cases, the real-time kinetic (RTK) receivers on the bicyclists and pedestrians achieve RTK-fixed or RTK-float accuracy, resulting in errors on the order of a few centimeters or a few decimeters, respectively; position accuracy on the instrumented interaction vehicles is on the order of 10 cm. We describe the data collection campaign at the University of Michigan's Mcity Test Facility for connected and automated vehicles, the interaction scenarios, and test conditions and show some visualizations of the test as well as initial evaluation results.

5. Dajiang Suo, Sarra Yako, Mathew Boesch, and Kyle Post, "Integrating STPA into ISO 26262 Process for Requirement Development," SAE Technical Paper 2017-01-0058, 2017, doi:10.4271/2017-01-0058.

    This paper describes a process map for integrating STPA into the functional safety process based on ISO 26262. Specifically, three steps in the process map are illustrated through a case study on an automotive system: (1) system assumptions and components from item definition are used to form the systems engineering foundations for STPA; (2) UCAs identified and safety constraints created in STPA Step 1 are used to evaluate existing safety goals with ASIL ratings developed from HARA; and (3) causal scenarios and factors for UCAs identified in STPA Step 2 help engineers create functional safety requirements and make architectural decisions. Developing requirements for automotive electrical/electronic systems is challenging, as those systems become increasingly SW-intensive. Designs must account for unintended interactions among SW features, combined with unforeseen environmental factors. In addition, engineers have to iteratively make architectural trade-offs and assign responsibilities to each component in the system to accommodate new safety requirements as they are revealed. STPA is a new technique for hazard analysis, in the sense that hazards are caused by unsafe interactions between components (including humans) as well as component failures and faults. Otherwise stated, STPA covers the safety analysis of system malfunctions as well as the SOTIF, in addition to functional safety. In particular, the paper illustrates how STPA can help evaluate safety and other system-level goals with ASIL classifications from ISO 26262's recommended HARA. The meta-model can be also used to provide guidance on making architectural decisions in order to create functional safety requirements. To make the process map applicable to different functional safety processes adopted by

OEMs, tool support is required. Guidelines on how to develop visualization tools based on the meta-model are given.

6. Oleg Lurie and Joseph Miller, "Hazard Analysis and Risk Assessment beyond ISO 26262: Management of Complexity via Restructuring of Risk-Generating Process," SAE Technical Paper 2018-01-1067, 2018, doi:10.4271/2018-01-1067.

The authors present an analytical model for an AEB example consisting of 20 operating states and 28 situations. The authors emphasize that all safety activities for AVs need to have their source in a HARA, encompassing all relevant aspects, including operational situations, description of functionality, and others. As part of a HARA analysis, a hidden semi-Markov chain is developed to help estimate the mean distance to accident (MDTA) which is used as a performance measure. The MDTA is calculated to be 153,000 km for all driving states based on the US driving statistics. This performance measure is useful for SOTIF validation. The parameter of controllability by the driver was additionally validated by Monte-Carlo simulation.

7. Philip Koopman and Michael Wagner, "Toward a Framework for Highly Automated Vehicle Safety Validation," SAE Technical Paper 2018-01-1071, 2018, doi:10.4271/2018-01-1071.

Safety validation is a crucial phase of AV safety. In this paper, the authors describe an approach to Highly Automated Vehicles (HAV) validation that includes the following elements: (a) a phased simulation and testing approach that emphasizes testing to mitigate residual validation risks from the previous phase while exploiting the speed vs. fidelity scalability properties inherent in testing and simulation; (b) observability points to produce human-interpretable data that both detect defect escapes from lower fidelity simulation phases and demonstrate the system is doing the right thing for the right reason; (c) explicit differentiation of the various roles of testing from checking for requirements gaps to checking for design faults and matching each type of testing with a relevant portion of a phased validation approach; and (d) a run-time monitoring approach to managing identified risks, catching assumption violations and unknown unknowns as they arise in fielded systems.

This approach can be expected to improve validation effectiveness compared to a brute-force testing campaign because it explicitly links testing and simulation activities to the risks being mitigated. This in turn permits concentrating effort on the sweet spot of defect detection for each particular level of simulation and test fidelity. The approach can also be expected to improve testing efficiency by concentrating each phase of testing on mitigating risks inherited from the preceding phase, without wasting resources revisiting low-risk conclusions or attempting to address out-of-scope risks that belong to other testing phases. While it is always better to ensure that all residual risks are known and mitigated to an acceptable level, it is clear that HAVs are going to be deployed even if there are places in which the safety argument contains risks that are not completely understood. The approach discussed in this paper provides a framework for establishing an initial safety argument based on multiple levels of simulation and testing fidelity. It also provides hooks for continuous improvement based on monitoring assumption violations and other residual validation risks during the course of testing and deployment.

8. Mario Berk, Hans-Martin Kroll, Olaf Schubert, Boris Buschardt, and Daniel Straub, "Bayesian Test Design for Reliability Assessments of Safety-Relevant Environment Sensors Considering Dependent Failures," SAE Technical Paper 2017-01-0050, 2017, doi:10.4271/2017-01-0050.

The perception of an AV is safety-critical; thus a correct assessment of the sensors' perception reliability is crucial for ensuring the safety of the automated driving functionalities. A Bayesian methodology for empirical reliability assessments of sensor-based environment perception is presented as an alternative to the commonly applied null hypothesis significance testing (NHST). It allows to estimate the necessary test drive effort to demonstrate the perception reliability of environment sensors, including dependent errors and time variable error probabilities. Furthermore, a solution to assess the reliability of a dependent redundant multi-sensor system is given. Applying the methodology in a case study shows that the empirical test drive effort may be unfeasibly large when the target level of safety is low. When working with a multi-sensor system in which the individual sensors are nearly independent of each other, the system's perception reliability is considerably higher than when utilizing a single sensor. This fact opens up the possibility of validating the perception reliability empirically with feasible test drive effort, when one is able to show that multiple sensors have a small error dependency. The verification of a small error dependency itself is however expected to require additional test drive efforts. Simplifications of the problem's complexity involve the treatment of different types of perception errors, the representation of the sensor data fusion with a majority voting scheme, and the approximation of the time-dependent performance of the perception induced through various physical influencing factors such as the weather.

9. Philip Koopman and Michael Wagner, "Challenges in Autonomous Vehicle Testing and Validation," *SAE Int. J. Trans. Safety* 4(1):2016, doi:10.4271/2016-01-0128.

There are significant challenges in the development of safe autonomous vehicles according to the V process, particularly during testing and validation. In this paper, the authors discuss three general approaches that seem promising: (a) phased deployment, (b) monitor/actuator architecture, and (c) fault injection. Phased deployment can be done by identifying well-specified operational concepts to limit the scope of operations and therefore the necessary scope of requirements. This would include limitations in environment, system health, and operational constraints that must be satisfied to enable autonomous operation. Validating that such operational constraints are enforced will be an essential part of ensuring safety and will have to show up in the V process as a set of operational requirements, validation, and potentially run-time enforcement mechanisms. For example, run-time monitoring might be required to monitor not only whether system state is in a permissible autonomy regime but also that assumptions made about the operational scenario in the safety argument are actually being satisfied and whether the system is actually in the operational scenario it thinks it is in. An aspect of restricted operational concepts that will require particular attention is ensuring that safety is maintained when an operational scenario suddenly becomes invalidated, due to, for example, an unexpected weather event or an infrastructure failure. A common approach that might help mitigate many of the challenges of autonomous vehicle safety is the use of a monitor/actuator architecture. A monitor/actuator architecture is one in which the primary functions are performed by one module (the actuator), and a paired module (the monitor) performs an acceptance test or other behavioral validation. If the actuator misbehaves, the monitor shuts the entire function down (both modules), resulting in a fail-silent system (i.e., any failure results in a silent component, sometimes also known as fail-stop, or fail-safe). Testing alone is infeasible to ensure ultra-dependable systems. Fault injection can play a useful role as part of a validation strategy that also includes traditional testing and non-test-based validation. This is especially true if fault injection is applied at multiple levels of abstraction rather than just at the level of stuck-at electrical connectors.

10. Philip Daian, Bhargava Manja, Grigore Rosu, Shinichi Shiraishi, and Akihito Iwai, "RV-ECU: Maximum Assurance In-Vehicle Safety Monitoring," SAE Technical Paper 2016-01-0126, 2016, doi:10.4271/2016-01-0126.

In this paper, the authors propose run-time verification ECU (RV-ECU) to separate safety from functionality and introduce a potential architecture for realizing such a practical separation. Although the application involves generic CAN communications, run-time verification can be applied with AVs. Run-time verification is a system analysis and approach that extracts information from the running system and uses it to assess satisfaction or violation of specified properties and constraints. RV-ECU uses run-time verification, a formal analysis subfield geared at validating and verifying systems as they run, to ensure that all manufacturer and third-party safety specifications are complied with during the operation of the vehicle. By compiling formal safety properties into code using a certifying compiler, the RV-ECU executes only provably correct code that checks for safety violations as the system runs. RV-ECU can also recover from violations of these properties, either by itself in simple cases or together with safe message-sending libraries implementable on third-party control units on the bus. Specifications checked at run-time can be both concise and formally precise, allowing for their development by engineers and managers not trained in formal methods while ensuring they are modular and easily sharable. The authors describe an implementation of such a system and demonstrate monitoring of safety specifications on the CAN bus.

I.8 **Conclusion**

SOTIF is an important safety category that is becoming well established in research and implementation and complements the categories of multi-agent and functional safety. SOTIF involves addressing hazards due to performance limitations or functional insufficiencies even in the case where there are no faults. There are two main sources for functional insufficiencies: (i) inherent uncertainty and complexity of the environment and (ii) intrinsic uncertainty within the functional implementation, for example, ML. In order to complete a thorough analysis of intrinsic uncertainties, the intended functionality has to be well understood and specified. Some possible sources of SOTIF hazards include under-specification, semantic gap, and deductive gap. In this chapter, we have characterized SOTIF and summarized its main concepts, particularly what is meant by functional insufficiencies and examples. In addition, an integrated framework to safety is presented which unifies the three safety categories (functional safety, SOTIF, and multi-user safety) using a risk-based approach. Furthermore, we summarized recent research on SOTIF safety measures and mechanisms to reduce risk. From a top-level perspective, there are several major approaches to reduce risk due to hazards and due to performance limitations or functional inefficiencies. One approach is to use rigorous systems engineering precepts to ensure the safety of AVs and to provide a strong safety assurance case. Another approach is to incorporate technical solutions, techniques, and methods (e.g., fault-tolerant mechanisms) from appropriate disciplines to achieve the same goal. We have summarized the work of Gauerhof et al. on the development of a set of well-structured validation targets to demonstrate that a ML-implemented pedestrian detection function fulfills its intended functionality. In terms of reducing risk due to hazards caused by under-specification, the authors suggest the following validation targets: (a) environment is sufficiently well known; (b) task is sufficiently well known; and (c) sensitivity against unpredictable or unspecified impact of environmental attributes is sufficiently low.

The use of appropriate standards (e.g., ISO 26262) or systems engineering concepts such as validation target cannot guarantee that hazards will not occur or that risk reduction efforts will be effective. We need to combine the above with technical measures or mechanisms that proactively deal with the hazards and help reduce risk to acceptable levels. In the context of the pedestrian detection function, analyzed by Gauerhof et al., the authors suggest some measures at the functional and system levels that will help reduce risk induced by functional insufficiencies. The suggested safety measures at the functional level include preprocessing of the ML input according to known factors that significantly influence performance and post-processing of the ML output to include adjustments of confidence levels based on factors known to influence performance, so that decisions about driving behavior and trajectory planning are adapted to the reliability of the perception function. The suggested safety measures at the system level include (a) diverse redundancy; (b) operating modes, also called degradation modes, which depend on the vehicle's environmental model; (c) transition between operating modes ensures a continuous driving behavior; (d) run-time monitoring of assumptions; and (e) established driver assistance systems (e.g., emergency brake assist). Pimentel and Bastian have suggested additional safety measures and mechanisms appropriate for reducing SOTIF risks.

References

1. Thrun, S. et al., "Stanley: The Robot that Won the DARPA Grand Challenge," *Journal of Field Robotics* 23, no. 9 (2006): 661–692.

2. Urmson, C. et al., "Autonomous Driving in Urban Environments: Boss and the Urban Challenge," *Journal of Field Robotics* 25, no. 8 (2008): 425–466.

3. Cheng, H., *Autonomous Intelligent Vehicles: Theory, Algorithms, and Implementation* (London: Springer, 2011).

4. Wang, F.-Y. et al., "IVS 05: New Developments and Research Trends for Intelligent Vehicles," *IEEE Intelligent Systems* 20, no. 4 (2005): 10–14.

5. Cheng, H. et al., "Interactive Road Situation Analysis for Driver Assistance and Safety Warning Systems: Framework and Algorithms," *IEEE Transactions on Intelligent Transportation Systems* 8, no. 1 (2007): 157–166.

6. International Organization for Standardization, "Road Vehicles – Functional Safety," ISO Standard 26262, 2011.

7. Koopman, P. and Wagner, M., "Autonomous Vehicle Safety: An Interdisciplinary Challenge," *IEEE Intelligent Transportation Systems Magazine* 9, no. 1 (2017): 90–96.

8. Avizienis, A. et al., "Basic Concepts and Taxonomy of Dependable and Secure Computing," *IEEE Transactions on Dependable and Secure Computing* 1, no. 1 (2004): 11–33.

9. Leveson, N.G., *Safeware: System Safety and Computers* (Addison-Wesley, 1995).

10. Wendorff, W., "Quantitative SOTIF Analysis for Highly Automated Driving Systems," *Safetronic, 2017 Conference Proceedings*, Stuttgart, Germany, 2017.

11. Thomas, J., Sgueglia, J., Suo, D., Leveson, N. et al., "An Integrated Approach to Requirements Development and Hazard Analysis," SAE Technical Paper 2015-01-0274, 2015, doi:10.4271/2015-01-0274.

12. Leveson, N.G., *Engineering a Safer World : Systems Thinking Applied to Safety* (MIT Press, 2012).

13. Young, W. and Leveson, N.G., "An Integrated Approach to Safety and Security Based on Systems Theory," *Communications of the Association for Computing Machinery (ACM)* 57, no. 2 (2014): 31–35.

14. Abdulkhaleq, A., Wagner, S., and Leveson, N., "A Comprehensive Safety Engineering Approach for Software-Intensive Systems Based on STPA," *3rd European STAMP Workshop, Conference Proceedings*, Amsterdam, The Netherlands, 2015.

15. Abdulkhaleq, A. et al., "A Systematic Approach Based on STPA for Developing a Dependable Architecture for Fully Automated Driving Vehicles," *4th European STAMP Workshop, Conference Proceedings*, Zurich, Switzerland, 2017.

16. Kopetz, H., *Real-Time Systems: Design Principles for Distributed Embedded Applications* (Kluwer Academic Publishers, 1997).

17. Koopman, P. and Wagner, M., "Toward a Framework for Highly Automated Vehicle Safety Validation," SAE Technical Paper 2018-01-1071, 2018, doi:10.4271/2018-01-1071.

18. International Electrotechnical Commission, "Functional Safety of Electrical/Electronic/Programmable Electronic Safety-Related Systems," IEC Standard 61508, 2010.

19. International Electrotechnical Commission, "Functional Safety – Safety Instrumented Systems for the Process Industry Sector," IEC Standard 61511, 2018.

20. Shalev-Shwartz, S., Shammah, S., and Shashua, A., "On a formal model of safe and scalable self-driving cars," *Computing Research Repository (CoRR)*, vol. arXiv:1708.06374 [cs.RO], 2017, [Online], http://arxiv.org/abs/1708.06374.

21. Gauerhof, L., Munk, P., and Burton, S., "Structuring Validation Targets of a Machine Learning Function Applied to Autonomous Driving," Gallina B. et al. (Eds.), *SAFECOMP 2018, LNCS 11093*, 2018, 45–58.

22. Feth, P., Adler, R., Fukuda, T., Ishigooka, T., Otsuka, S., Schneider, D., Uecker, D., and Yoshimura, K., "Multi-Aspect Safety Engineering for Highly Automated Driving Looking Beyond Functional Safety and Established Standards and Methodologies," Gallina B. et al. (Eds.), *SAFECOMP 2018, LNCS 11093*, 2018, 59–72.

23. Pimentel, J. and Bastiaan, J., "Characterizing the Safety of Self-Driving Vehicles: A Fault Containment Protocol for Functionality Involving Vehicle Detection," *2018 IEEE International Conference on Vehicular Electronics and Safety (ICVES)*, Madrid, Spain, September 12-14, 2018.

24. Pimentel, J., Bastiaan, J., and Zadeh, M., "Numerical Evaluation of the Safety of Self-Driving Vehicles: Functionality Involving Vehicle Detection," *2018 IEEE International Conference on Vehicular Electronics and Safety (ICVES)*, Madrid, Spain, September 12-14, 2018.

25. Schorn, C., Guntoro, A., and Ascheid, G., "Efficient On-Line Error Detection and Mitigation for Deep Neural Network Accelerators," Gallina, B. et al. (Eds.), *SAFECOMP 2018, LNCS 11093*, 2018, 205–219.

26. International Organization for Standardization, ISO/WD PAS 21448, "Road Vehicles - Safety of the Intended Functionality," ISO Working Draft, 2013.

27. Burton, S., Gauerhof, L., and Heinzemann, C., "Making the Case for Safety of Machine Learning in Highly Automated Driving," *International Conference on Computer Safety, Reliability, and Security (SAFECOMP), Conference Proceedings*, Trento, Italy, 2017.

28. Salay, R., Queiroz, R., and Czarnecki, K., "An Analysis of ISO 26262: Machine Learning and Safety in Automotive Software," SAE Technical Paper 2018-01-1075, 2018, doi:10.4271/2018-01-1075.

29. Geiger, A. et al., "Vision Meets Robotics: The KITTI Dataset," *International Journal of Robotics Research* 32, no. 11 (2013): 1231–1237.

30. Bi, X. et al., "A New Method of Target Detection Based on Autonomous Radar and Camera Data Fusion, SAE Technical Paper 2017-01-1977, 2017, doi:10.4271/2017-01-1977.

31. Realpe, M., Vintimilla, B., and Vlacic, L., "Towards Fault Tolerant Perception for Autonomous Vehicles: Local Fusion," *7th IEEE International Conference on Robotics, Automation and Mechatronics (RAM), Conference Proceedings*, Angkor Wat, Cambodia, 2015.

Fault-Tolerant Ability Testing for Automotive Ethernet

Jinpeng Xu and Feng Luo
Tongji University

With the introduction of BroadR-Reach and time-sensitive networking (TSN), Ethernet has become an option for in-vehicle networks (IVNs). Although it has been used in the IT field for decades, it is a new technology for automotive, and thus requires extensive testing. Current test solutions usually target specifications rather than the in-vehicle environment, which means that some properties are still uncertain for in-vehicle usage (e.g., fault tolerance for shorted or open wires). However, these characteristics must be cleared before applying Ethernet in IVNs, because of stringent vehicular safety requirements. Because CAN is usually used for these environments, automotive Ethernet is expected to have the same or better level of fault tolerance.

Both CAN and BroadR-Reach use a single pair of twisted wires for physical media; thus, the traditional fault-tolerance test method can be applied for automotive Ethernet. However, because of the new transceiver circuit, some test methods need to be modified. This paper analyzes and tests the physical communication mechanism for BroadR-Reach to compare with the CAN bus. Test cases, such as a shorting a wire to BAT or GND, ground shift, and load capacity, are applied to TJA1100 from NXP. Although these test results are related to chip manufacturers and production batches, the analysis results still have guiding significance for the application of automotive Ethernet to avoid dangerous conditions.

CITATION: Xu, J. and Luo, F., "Fault-Tolerant Ability Testing for Automotive Ethernet," SAE Technical Paper 2018-01-0755, 2018, doi:10.4271/2018-01-0755.

Introduction

In-vehicle electronic systems are rapidly developing in complexity and diversity, including autonomous driving and advanced driver-assistance systems. Both of these require more bandwidth than can be afforded by traditional IVNs. Based on a comparison with USB, LVDS, fire wire and other buses, Ethernet with BroadR-Reach [1] seems to be the right choice because of its high bandwidth and electromagnetic compatibility (EMC) performance [2]. With the introduction of audio-video bridging (AVB) [3] and TSN [4], automotive Ethernet has the potential to be an IVN during runtime because of its deterministic media access control (MAC) services to the upper layer. The real-time ability [5, 6, 7] and fault-tolerance topology [8, 9] show that it is possible to deploy Ethernet-based backbones that consolidate all automotive domains on a single physical layer. This is attractive, as Ethernet has been used in the IT field for decades, and many resources are reusable in future developments.

Although Ethernet has been used for vehicle diagnostics and entertainment systems, it is risky to replace CAN because of Ethernet's stricter requirements. As a new technology for the automotive field, extensive testing is needed. AUTOSAR, OPEN SIG, Avnu, and other organizations have defined the relevant test specifications. These test cases cover the physical layer, data link layer, network communication protocols, and performance on electronic control unit (ECU) or switches, as shown in Figure 1.

In order to overcome the challenges for automotive Ethernet testing, companies from the IT (e.g., Ixia, Spirent), automotive (e.g., Vector, Elektrobit), and measurement fields (e.g., Keysight, Tektronix) have provided tools, such as physical signal measurement, network emulators, sniffers, and clock measurement tools, to cover the test requirements. Most of these tools are for the performance and conformance of Ethernet, which is of interest for entertainment systems. With the upcoming of TSN, the application of Ethernet for control- related subnets and backbones is possible, which leads to higher

FIGURE 1 Automotive ethernet testing.

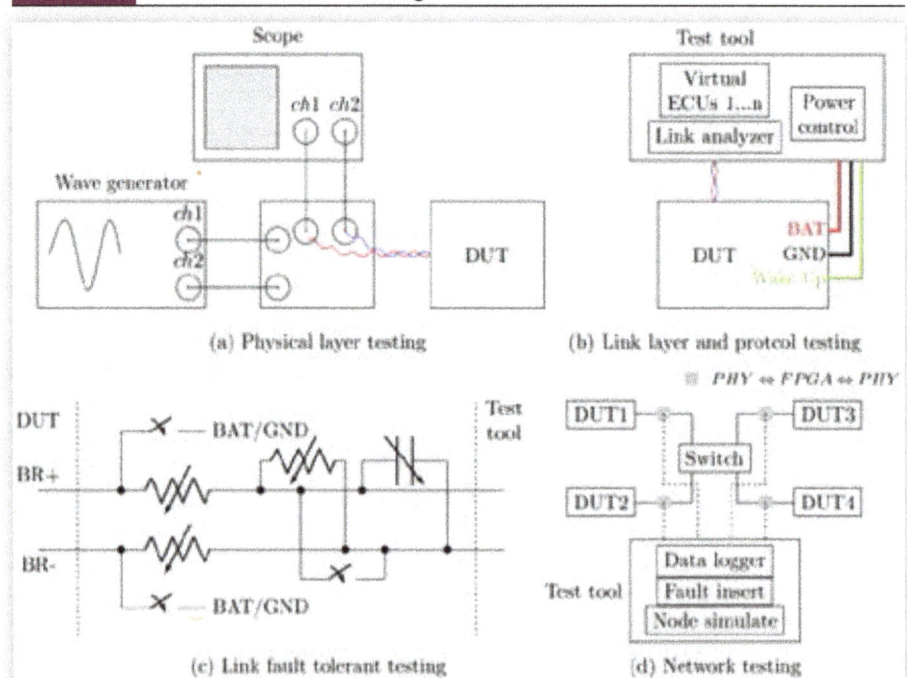

(a) Physical layer testing

(b) Link layer and protcol testing

(c) Link fault tolerant testing

(d) Network testing

requirements for fault tolerance. Both CAN and BroadR-Reach use one pair of twisted wires for physical media, it means the traditional fault-tolerant test method can still apply for automotive Ethernet as shown in Figure 1 (c). The test methods in this paper come from fault tolerant mode test of GMW14241. However, due to the new transceiver circuit, some test methods need to be modified as required.

This paper aims to test the fault tolerance of automotive Ethernet and compare it with that of a CAN bus. Section 2 analyzes the BroadR-Reach physical link and makes some assumptions regarding the fault-tolerance performance. Section 3 introduces the test methods and results to show the fault tolerance. Section 4 analyzes these results, points out causes for the results, and puts forward some suggestions for improvement. Section 5 summarizes the results and provides some advice for the use of automotive Ethernet to avoid dangerous conditions.

Physical Layer Analysis

BroadR-Reach technology began with the IEEE 802.3 1000BASE-T standard; thus, some basic principles are reused, as shown in Figure 2. The 1000BASE-T standard uses five voltage values: –2, –1, 0, 1, and 2 on one pair of wires; thus, four pairs wires allow $5^4 =$ 625 possible code words. 256 of these are data words, and the reset are used for redundancy and control (4D-PAM5) [10].

BroadR-Reach technology, like other 100BASE Ethernet, interfaces with MAC via a standard media-independent interface (MII), which passes 4 bits of data on each clock and runs at 25 MHz. The next transmission (or reversed reception) step is a 4B/3B clock conversion performed by the physical coding sublayer (PCS). The 4-bit blocks are converted to 3-bit blocks at a clock rate of $33.\dot{3}$ MHz. Rather than translating the 3-bit block directly to the lower layer, a technique called side-stream scrambling is applied. This step improves DC balance and avoids problems with clock synchronization caused by long strings of 0 s or 1 s. As the upper right of Figure 2 shows, although no frame is transmitting on the link, the differential signal still produces enough jump for synchronization.

Finally, each 3-bit block is encoded as a pair of ternary symbols with values −1, 0, and +1 (3B/2 T). Because 3 bits means 8 possible values, 2 ternary symbols are used for encoding. As Figure 3 shows, the bit map of BroadR-Reach during tx_mode is SEND_N. These ternary symbols are transmitted on one pair of wires at 66.6 MHz.

BroadR-Reach is a full-duplex transceiver on one pair of wires. This is achieved by a hybrid inside physical layer chip. Figure 4 shows an example for 1000BASE-T, where

FIGURE 2 1000BASE-T and BroadR-Reach.

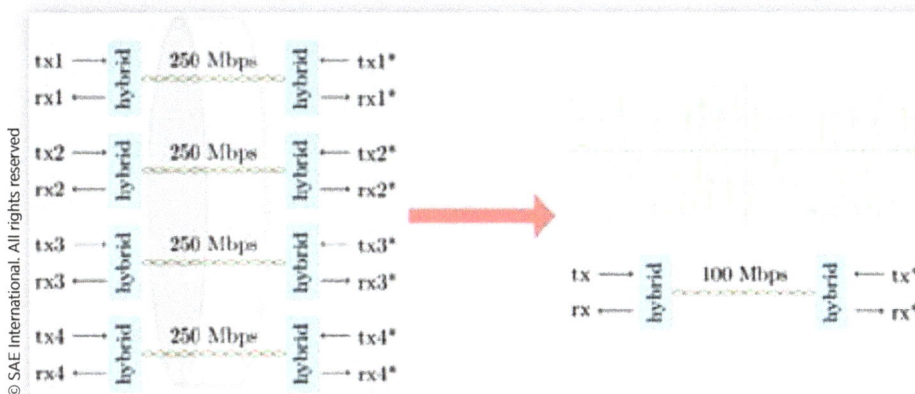

FIGURE 3   3B/2 T for BroadR-Reach.

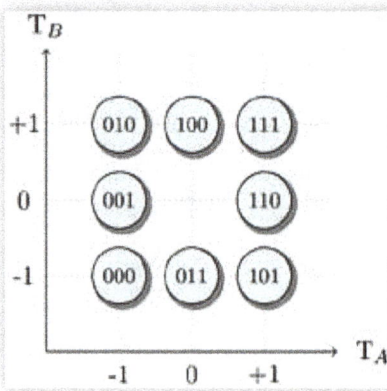

transmit signal A travels through two paths to ground. The received signal B also goes through R4 with the help of a network transformer; thus, A and B are mixed at R4. With the subtraction at receiver side, the theoretical receiver path is completely decoupled from the transmit path. Because the on-board equipment is powered by batteries, there is no need for high-voltage isolation, and the AC coupling capacitance is adopted instead of the transformer at the output end, as shown in Figure 5.

This hybrid structure allows only two nodes on one link, which means that automotive Ethernet is point-to-point communication, and multi-nodes must swap frames by an Ethernet switch. This characteristic may increase cost, but results in better signal quality. The signal waveform of the bus network system is usually worse (e.g., CAN signal ringing) with growing node numbers and link length of the trunk or branch. However, automotive Ethernet is not concerned with this problem, as each link is physically separate from the others. A fault on one link will not lead to unpredictable failures on other links.

FIGURE 4   Hybrid structure [11] for 1000BASE-T.

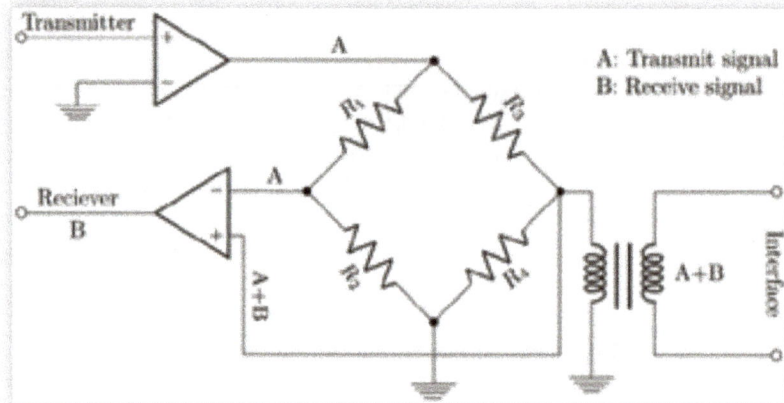

FIGURE 5   Testing interface circuit for TJA1100.

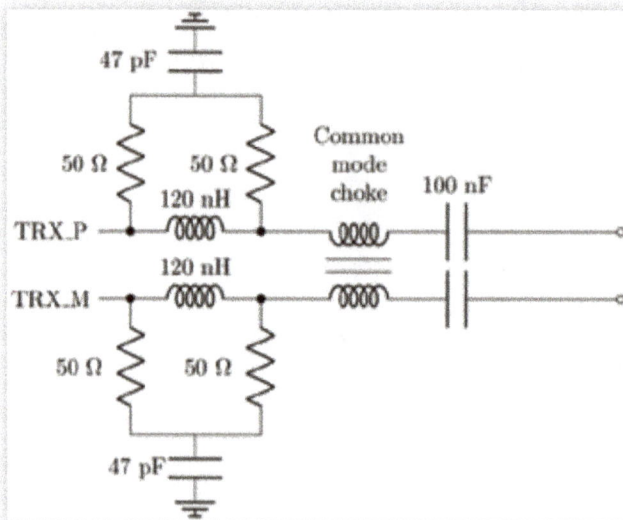

However, Ethernet still has a disadvantage compared with CAN. CAN supports error detection, as [12] defined. Corrupted frames are checked by the transmitting node and any normally operating receiving node. The automatic retransmission guarantees that the frames are correctly transmitted as far as possible. Ethernet does not have this mechanism, and the transmission failure is only known by the receiving node via the rx_err signal of the MII or CRC check. If the receiver's software ignores this error, the frame will be lost without any notification to the transmitting node. Transmission control protocol (TCP) can be used to help retransmission, but the real-time ability is unacceptable for control frames. Seamless redundant technology in TSN could reduce packet loss while ensuring real-time ability, but it still cannot avoid the packet loss. Therefore, it is of practical significance to study the fault tolerance of BroadR-Reach links. Although the real circuit inside the chip is unknown, below are some assumptions according to Figures 4 and 5.

a. Shorting one wire to BAT or GND does not affect communication due to the AC coupling capacitance;
b. Ground shift does not affect communication for the same reason;
c. The influence of parallel capacitance on the twisted wires will increase as the signal frequency increases compared with CAN;
d. Communication cannot continue with only one wire connected.

Fault Tolerance Testing

As connectors and wires age, several faults may appear, such as external resistance and capacitance, open cables, short circuits and ground shift. In the application of Ethernet for IVNs with CAN, the same or better level of fault tolerance is expected. This section will test these abilities on Ethernet and CAN. The transceiver used here is TJA1100 for Ethernet and TJA1040 for CAN at 500 kps with two nodes, as shown in Figure 6. The BAT voltage is 12 V.

Wire Short or Open Testing

Wire a short or break is possible during runtime. Usually the network is not required to continue normal communication, but the component must not be damaged during the fault. The test results are shown below.

A good result for Ethernet is when one wire is shorted to GND. No matter which wire is shorted to GND, the communication is not effected. This result is better than that of the same fault in CAN. However, an unexpected result occurs when one wire is shorted to BAT. Ethernet is expected to be unaffected by this fault because of the AC

FIGURE 6 Basic testing circuit.

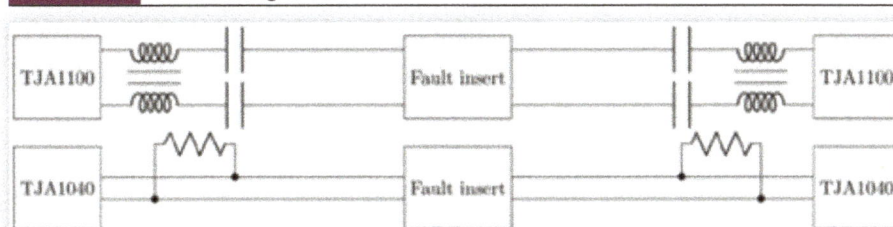

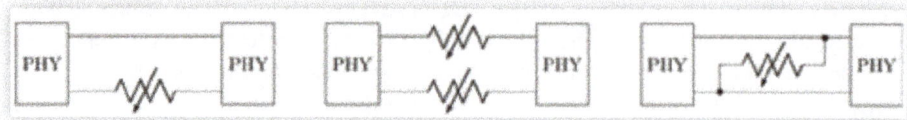

coupling capacitance; however, the real test result shows the damaged node cannot recover after a power cycle, which means this is a permanent fault. The reason for this damage will be discussed in the results analysis section.

Resistance Testing

The resistance is applied to one or both wires, as Figure 7 shows. This test case can be used to show the fault tolerance under poor connection conditions. With varying resistance, the CAN fault is in the form of an error frame. For Ethernet, the fault means the establishment of the link fails or the frame is lost.

The test results are summarized below. For each test case, the left value indicates that the network could work correctly under this condition. For the right value, the network would begin to have errors.

A wire pulled down to GND is not tested. One wire pulled down to GND would have no effect, according to Table 1. If both wires are pulled down to GND, the fault can be reduced to the case of parallel resistance, which is already tested.

A wire pulled up to BAT with resistance will be tested when analyzing the reason for short-circuit damage.

Capacitance Testing

The capacitance is parallel on twisted wires, as Figure 8 shows. This test case is also used to show the fault tolerance under poor connections. Due to the higher physical signal rate (66.6 MHz for Ethernet and 500 kps for CAN), the parallel capacitance is expected to have a greater effect on Ethernet.

Blew is the results for Ethernet and CAN.

- Ethernet could continue communication under 100 pF with additional startup delay;

- Ethernet stops communication under 220 pF;

- CAN could continue communication under 10 nF;

- CAN stops communication under 22 nF

The results show that Ethernet is sensitive to capacitance. It is easy to reach 100 pF by increasing the wire length, which means a link with several meters may fail to start up in time.

TABLE 1 Testing result for Ethernet

Fault type	During fault	After fault remove
One wire short to GND	No fault	No damage
One wire short to BAT	Communication stop	Damaged
One wire open	Communication stop	No damage
Short between twisted wires	Communication stop	No damage

TABLE 2 Testing result for CAN

Fault type	During fault	After fault remove
One wire short to GND	CANH: Communication stop CANL: No fault	No damage
One wire short to BAT	CANH: No fault. CANL: Communication stop	No damage
One wire open	Communication stop	No damage
Short between twisted wires	Communication stop	No damage

TABLE 3 Testing result for resistance testing

Network type	Asymmetric resistance (Ω)		Symmetric resistance (Ω)		Parallel resistance (Ω)	
CAN	160	242	121	141	20	10
Ethernet	59	82	30	39	180	170

TABLE 4 Testing result for ground shift

Ground shift type	Voltage	result
Constant voltage	0 V-10 V	No effect
Sine voltage	−5 V-5 V @ (1 kHz, 10 kHz, 50 khz)	No effect
Pulse voltage	0 V to 10 V	Frame lost during voltage change. Communication continue after pulse.
Square voltage	0 V to 10 V @ 1 Hz	Frame lost during voltage change. Communication stop after few seconds Communication recovery after reboot

Ground Shift Testing

The traditional ground shift requirement for CAN is ±2 V without error frames. Due to the coupling capacitance, a constant voltage ground shift is not expected to have an effect on Ethernet; thus, the AC ground shift ability is tested as shown in Figure 9.

The test results show that constant ground and AC ground shift have no effect on communication. The pulse voltage leads to frame loss during the voltage change. With the square wave, communication stops after a few seconds. However, these faults are temporary, and the node recovers after reboot.

Result Analysis

The testing results can be summarized as follows:

a. Ethernet requires better connections;
b. Ethernet has better resistance to ground shift;
c. Ethernet wires should avoid short circuits with high-voltage pins.

It is easy to accept a) as it may be unavoidable with increasing physical frequency. However, this risk should

FIGURE 8 Capacitance testing setup.

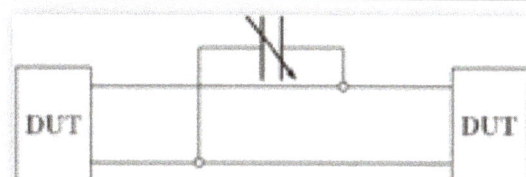

FIGURE 9 Ground shift testing setup.

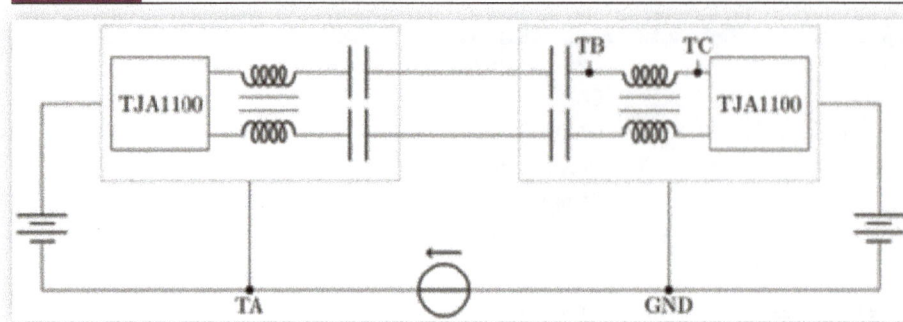

be taken into account to assess whether the link in the product life cycle can be guaranteed. In order to detect the link fault and recovery after faults, a link management system, such as a bus-off recovery for CAN, is required for Ethernet to correctly restart the transceiver and record the fault state.

Result b) is useful for the resistance to common-mode disturbances. In Figure 10, the blue line (5 V/div) stands for TA in Figure 9, the purple line (2 V/div) stands for TB, and the yellow line (2 V/div) indicates TC. The voltage decrease between TB and TC shows the benefit of a common-mode choke. However, the common-mode disturbance still has an effect on the nodes, as the left side of Figure 10 shows. The sine wave is superimposed onto the receiving port. Because Ethernet uses differential signal transmission, the communication continues correctly. In the case of pulse- or square-wave interference, the ability of common-mode suppression is limited. Although the inhibition still takes effect, the link loses frames or breaks down during the fault.

Result c) is not expected, as Ethernet has coupling capacitance and supports power over data lines (PoDL). However, the test result shows that this expectation is incorrect. As Figure 11 shows, the circuits for PoDL and the fault are different. Additional inductances are used to protect the link for PoDL. If the voltage on TA is slowly increased

FIGURE 10 Testing wave for sine and pulse voltage.

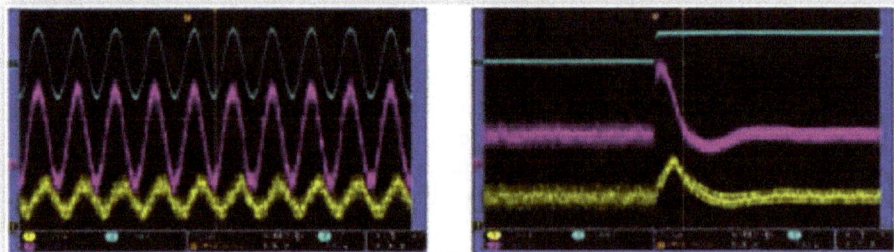

FIGURE 11 PoDL and fault setup.

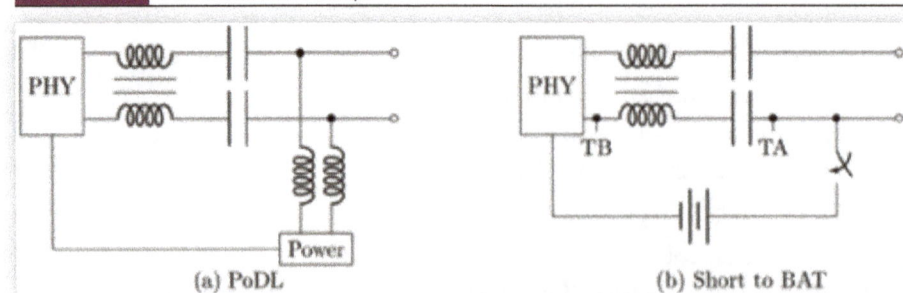

(a) PoDL   (b) Short to BAT

from 1 V to BAT (less than 0.5 V/s during the test), the communication is not affected, and there is no damage to the transceiver.

The data sheet for TJA1100 shows that the max voltage on TRX_P/M is 4.6 V, which is easily achieved when the short occurs, as shown in Figure 12. The red line (5 V/div) shows the TRX_P voltage (TA in Figure 11) and the blue line (5 V/div shows the related voltage of TRX_P on the chip side (TB). The left image shows the short-term fault; it can be seen that the voltage on the chip side rises, but does not reach the limit. Thus, communication continues. However, the long-term fault leads voltage on the chip side to rise above 5 V, and communication stops after about 2.2 μs.

The long-term fault (short point close to PHY_1) leads to an unstable physical-layer chip state. After the fault, two nodes are connected to the normal node, and reboot 60 times to show their status. Table 5 shows the results for two tests. PHY_1 cannot start communication or recover lost frames after reboot. This means that if this fault occurs during runtime, the communication of the vehicle become unstable.

This is even worse than a complete lack of communication, as the unstable condition increases the difficulty of troubleshooting.

The test results for PHY_2 are better than those for PHY_1. The reason may be the 2-m cable between the two nodes, which provides additional protection between the fault point and PHY_2. In order to show the effect of additional resistance, the test case shown on the left of Figure 13 is applied. The results show that the frame is lost when the fault occurs and R1 is below 330 Ω; however, the communication continues and the nodes are not damaged when R1 ≥ 0.75 Ω. This means that protecting the pins is not difficult. Compared with the design reference from NXP, the circuit in Figure 5 does not have any electrostatic discharge (ESD) components. Although such components are usually targeted for voltages at the kilovolt level, they can be used to prevent this damage, as Figure 13 (b) shows. When the voltage rises above VCC plus the diode lead voltage, the diode is open

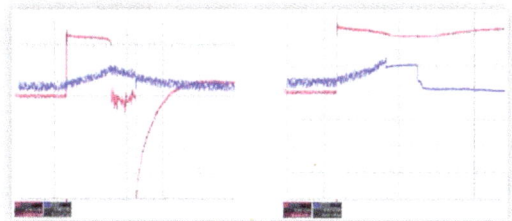

FIGURE 12 Short TRX M with BAT (2us/div).

TABLE 5 60 reboot after short with BAT at PHY_1

	1st test		2st test	
Fault type	PHY_1	PHY_2	PHY_1	PHY_2
No communication	21	0	16	0
Link startup delay and frame lost during communication	9	0	14	1
Frame lost during communication	30	0	30	1
Link startup delay	0	0	0	1
No fault	0	60	0	57

FIGURE 13 Short testing with additional resistance.

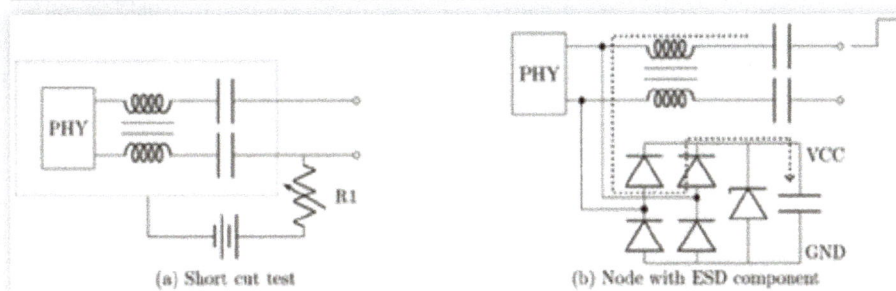

(a) Short cut test   (b) Node with ESD component

FIGURE 14 Short testing with ESD component.

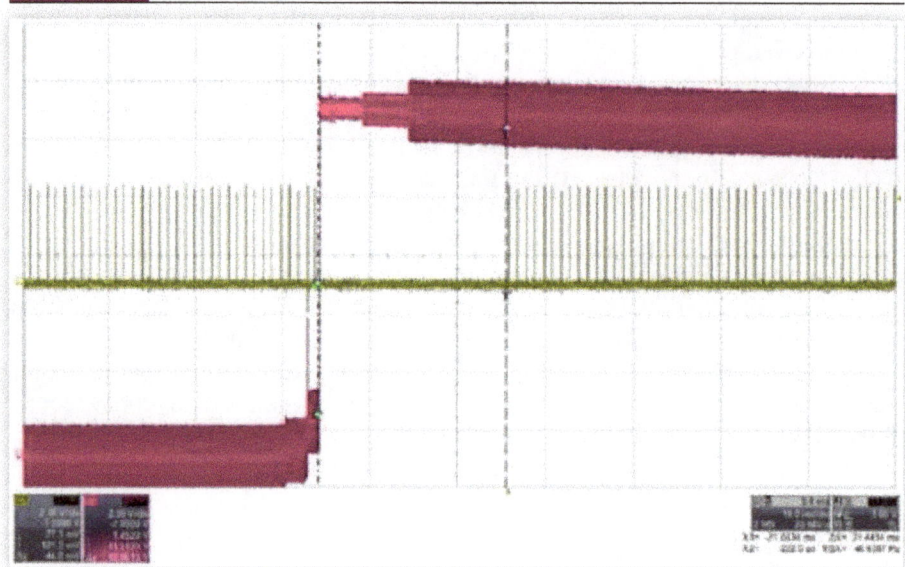

and the pulse energy is absorbed by the capacitor. The real test result is shown in Figure 14, where the red line is TRX_M and the yellow line is RX_DV of MII. This means that communication stops during the fault and recovers after about 21 ms.

Summary/Conclusions

In this study, the fault tolerance for automotive Ethernet was tested. The results show that Ethernet has less tolerance for connector and wire aging, which means better components are required to guarantee their physical properties during the life cycle of the system. The impact of vibration was not assessed in this paper, but should not be ignored as the short-term open connection, which is unimportant for CAN, may lead to Ethernet packet loss. Considering the high bandwidth performance of Ethernet, the cost for higher-quality connectors and wires is well deserved.

With the help of ESD components, Ethernet shows better tolerance for faults when a wire shorted to BAT. Although some time is needed for recovery from the fault, a continuous short-circuit state does not result in long communication failures. The lost frames can be retransmitted by the application, or transmitted with a redundant link, which ensures that the failure has a minimal impact on communication.

Other faults may lead to frame loss or stopped communication, but not permanent damage. Each result shows equal or better fault tolerance compared with CAN, which means automotive Ethernet has the potential to be more widely used with better connections and improved circuits.

Contact Information

Jinpeng Xu
Tongji University, Shanghai, China
Tel: +86-021-69589482
1310790@tongji.edu.cn

Feng Luo
Tongji University, Shanghai, China
Tel: +86-021-69583892
1310790@tongji.edu.cn

Acknowledgments

This work was financially supported by Shanghai Automotive Industry Science and Technology Development Foundation (1515).

Definitions/Abbreviations

IVN - In-vehicle network
TSN - Time-sensitive networking
AVB - Audio Video Bridging
MAC - Media access control
CAN - Controller area network
PCS - Physical coding sublayer
TCP - transmission control protocol

References

1. BMW, Broadcom, NXP, and ..., "OPEN Alliance SIG," 2011, accessed December 24, http://www.opensig.org/.
2. Matheus, K. and Königseder, T., "Automotive Ethernet," 2015.
3. IEEE, "Audio Video Bridging Task Group," 2011, accessed January 2, http://www.ieee802.org/1/pages/avbridges.html.
4. IEEE, "Time-Sensitive Networking Task Group," 2012, accessed January 2, http://www.ieee802.org/1/pages/tsn.html.
5. Park, C., Lee, J., Tan, T., and Park, S., "Simulation of Scheduled Traffic for the IEEE 802.1 Time Sensitive Networking," *Information Science and Applications (ICISA) 2016*, Springer, 2016, 75-83.
6. Craciunas, S.S., Oliver, R.S., Chmelík, M., and Steiner, W., "Scheduling Real-Time Communication in IEEE 802.1 Qbv Time Sensitive Networks," *Proceedings of the 24th International Conference on Real-Time Networks and Systems*, 2016.
7. Steinbach, T., Lim, H.T., Korf, F., Schmidt, T.C. et al., "Beware of the Hidden! How Cross-Traffic Affects Quality Assurances of Competing Real-Time Ethernet Standards for In-Car Communication," *2015 IEEE 40th Conference on Local Computer Networks (LCN)*, October 26-29, 2015.
8. Farzaneh, M.H. and Knoll, A., "An Ontology-Based Plug- And-Play Approach for in-Vehicle Time-Sensitive Networking (TSN)," *7th IEEE Annual Information Technology, Electronics and Mobile Communication Conference, IEEE IEMCON 2016*, Canada, Vancouver, BC, October 13–15, 2016.

9. Kehrer, S., Kleineberg, O., and Heffernan, D., "A Comparison of Fault-Tolerance Concepts for IEEE 802.1 Time Sensitive Networks (TSN)," *19th IEEE International Conference on Emerging Technologies and Factory Automation, ETFA 2014*, Barcelona, Spain, September 16-19, 2014.

10. Patwardhan, S., "Gigabit Ethernet over Copper: Hardware Architecture and Operation," DELL, 2001, accessed December 25, http://www.dell.com/content/topics/global.aspx/power/en/ps4q01_patward?c=u.

11. Kauffels, F.-J., "1000 BASE-T: Gigabit Ethernet über Kupferverkabelung," 2012, accessed December 25, http://www.comconsult-research.de/1000-base-t-gigabit-ethernet-uber-kupferverkabelung-12/.

12. ISO, *ISO 11898-1:2003-Road Vehicles-Controller Area Network* (Geneva, Switzerland: International Organization for Standardization, 2003).

An Analysis of ISO 26262: Machine Learning and Safety in Automotive Software

Rick Salay, Rodrigo Queiroz, and Krzysztof Czarnecki
University of Waterloo

Machine learning (ML) plays an ever-increasing role in advanced automotive functionality for driver assistance and autonomous operation; however, its adequacy from the perspective of safety certification remains controversial. In this paper, we analyze the impacts that the use of ML within software has on the ISO 26262 safety lifecycle and ask what could be done to address them. We then provide a set of recommendations on how to adapt the standard to better accommodate ML.

CITATION: Salay, R., Queiroz, R., and Czarnecki, K., "An Analysis of ISO 26262: Machine Learning and Safety in Automotive Software," SAE Technical Paper 2018-01-1075, 2018, doi:10.4271/2018-01-1075.

Introduction

The use of machine learning (ML) is on the rise in many sectors of software development, and automotive software development is no different. In particular, Advanced Driver Assistance Systems (ADAS) and Automated Driving Systems (ADS) are two areas where ML plays a significant role [1, 2]. In automotive development, safety is a critical objective, and the emergence of standards such as ISO 26262 [3] has helped focus industry practices to address safety in a systematic and consistent way. Unfortunately, ISO 26262 was not designed to accommodate technologies such as ML, and this has created a tension between the need to innovate and the need to improve safety.

In response to this issue, research has been active in several areas. Recently, the safety of ML approaches in general have been analyzed both from theoretical [4] and pragmatic perspectives [5]. However, most research is specifically about neural networks (NN). Work on supporting the verification & validation (V&V) of NNs emerged in the 1990's with a focus on making their internal structure easier to assess by extracting representations that are more understandable [6]. General V&V methodologies for NNs have also been proposed [7, 8]. More recently, with the popularity of deep neural networks (DNN), verification research has included more diverse topics such as generating explanations of DNN predictions [9], improving the stability of classification [10] and property checking of DNNs [11].

Despite their challenges, NNs are already used in high assurance systems (see [12] for a survey), and safety certification of NNs has received some attention. Pullum et al. [13] give detailed guidance on V&V as well as other aspects of safety assessment such as *hazard analysis* with a focus on adaptive systems in the aerospace domain. Bedford et al. [14] define general requirements for addressing NNs in any safety standard. Kurd et al. [15] have established criteria for NNs to use in a safety case.

The recent surge of interest in ADSs has also been driving research in certification. Koopman and Wagner [2] identify some of the key challenges to certification, including ML. Martin et al. [16] analyze the adequacy of ISO 26262 for an ADS but focuses on the impact of the increased complexity it creates rather than specifically the use of ML. Spanfelner et al. [1] assess ISO 26262 from the perspective of driver assistance systems. Finally, Burton et al. [17] explore the kind of safety case that is required for an ADS that uses ML components.

The contribution of the current paper is complementary to the above research. We analyze the impact that the use of ML-based software has on various parts of ISO 26262. Specifically, we consider its impact in the areas of hazard analysis and in the phases of the software development process. In all, we identify five distinct problems that the use of ML creates and make recommendations on steps toward addressing these problems both through changes to the standard and through additional research.

The remainder of the paper is structured as follows. In the next section we give the required background on ISO 26262 and ML. Following this is the analysis of the ISO 26262 safety lifecycle with five subsections describing each impacted area and the corresponding recommendations. Finally, we summarize and give concluding remarks.

Background

ISO 26262

ISO 26262 is a standard that regulates functional safety of road vehicles. It recommends the use of a Hazard Analysis and Risk Assessment (HARA) method to identify hazardous

FIGURE 1 ISO 26262 part 6 - Product development at the software level.

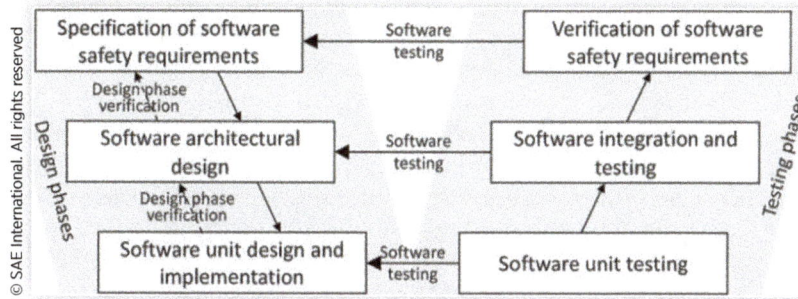

events in the system and to specify safety goals that mitigate the hazards. The standard has 10 parts, but we focus on Part 6: "product development at the software level". The standard follows the well-known V model for engineering shown in Figure 1.

An Automotive Safety Integrity Level (ASIL) refers to a risk classification scheme defined in ISO 26262 for an item (e.g., subsystem) in an automotive system. The ASIL represents the degree of rigor required (e.g., testing techniques, types of documentation required, etc.) to reduce the risk of the item, where ASIL D represents the highest and ASIL A the lowest risk. If an element is assigned QM (Quality Management), it does not require safety management. The ASIL assessed for a given hazard is first assigned to the safety goal set to address the hazard and is then inherited by the safety requirements derived from that goal.

Part 6 of the standard specifies the compliance requirements for software development. For example, Figure 2 shows the error handling mechanisms recommended for use as part of the architectural design. The degree of recommendation for a method depends on the ASIL and is categorized as follows: ++ indicates that the method is highly recommended for the ASIL; + indicates that the method is recommended for the ASIL; and o indicates that the method has no recommendation for or against its usage for the ASIL. For example, *Graceful Degradation* (1b) is the only highly recommended mechanism for an ASIL C item, while an ASIL D item would also require *Independent Parallel Redundancy* (1c).

Machine Learning

In this paper, we are concerned with software implementation using ML. We call a *programmed component* to be one that is implemented using a programming language, regardless of whether the programming was done manually or automatically (e.g., via code generation). In contrast, an *ML component* is one that is a trained model using a supervised, unsupervised or reinforcement learning (RL) approach.

An ML component can be trained offline during system development or online in a running system. For ML components in automotive systems, we assume that online learning is limited to non safety-critical functionality. For example, an ML component could be trained online to learn a driver's infotainment preferences. The key weakness of online learning with respect to functional safety assurance is that a safety argument cannot be made and assessed ahead of time. Thus, for the applications of ML discussed in this paper, we assume training is done offline.

FIGURE 2 ISO 26262 Part 6 - Mechanisms for error handling at the soft- ware architectural level.

Methods		ASIL			
		A	B	C	D
1a	Static recovery mechanism	+	+	+	+
1b	Graceful degradation	+	+	++	++
1c	Independent parallel redundancy	o	o	+	++
1d	Correcting codes for data	+	+	+	+

There are several characteristics of ML that can impact safety or safety assessment.

Non-transparency. All types of ML models contain knowledge in an encoded form, but this encoding is more *transparent* - i.e., easier to interpret by humans - in some types than others. Bayesian Networks are transparent since the nodes are random variables and can represent human-defined concepts. For example, a Bayesian Network for weather prediction may have nodes such as "precipitation type", "temperature", etc. In contrast, NN models are considered non-transparent and significant research effort has been devoted to making them more transparent (e.g., [6, 9]). Increasing ML model expressive power is typically at the expense of transparency but some research efforts focus on mitigating this [18]. Non-transparency is an obstacle to safety assurance because it is more difficult for an assessor to develop confidence that the model is operating as intended.

Error rate. An ML model typically does not operate perfectly and exhibits some error rate. Thus, "correctness" of an ML component, even with respect to test data, is seldom achieved and it must be assumed that it will periodically fail. Furthermore, although an estimate of the true error rate is an output of the ML development process, there is only a statistical guarantee about the reliability of this estimate. Finally, even if the estimate of the true error rate was accurate, it may not reflect the error rate the system actually experiences while in operation after a finite set of inputs because the true error is based on an infinite set of samples [4]. These characteristics must be considered when designing safe system using ML components.

Training-based. Supervised and unsupervised learning based ML models are trained using a subset of possible inputs that could be encountered operationally. Thus, the training set is necessarily incomplete and there is no guarantee that it is even representative of the space of possible inputs. In addition, learning may overfit a model by capturing details incidental to the training set or training environment rather than general to all possible inputs in the operational environment [17]. RL suffers from similar limitations since it typically explores only a subset of possible behaviours during training. The uncertainty that this creates about how an ML component will behave is a threat to safety. Another factor is that, even if the training set is representative, it may under-represent the safety-critical cases because these are often rarer in the input space [4]. Finally, over time, the underlying distribution of inputs in the operational environment may drift from that of the training set, degrading safety [17].

Instability. More powerful ML models (e.g., DNN) are typically trained using local optimization algorithms, and there can be multiple optima. Thus, even when the training set remains the same, the training process may produce a different result. However, changing the training set also may change the optima. In general, different optima may be far apart structurally, even if they are similar behaviourally. This characteristic makes it difficult to debug models or reuse parts of previous safety assessments.

Analysis of ISO 26262

In this section, we detail our analysis of ML impacts on ISO 26262. Since ML-based software is a specialized type of software, we classify an area of the standard as *impacted* when it is relevant to software and the treatment of ML-based software should differ from the existing treatment of software by the standard. Applying this criterion to the ten parts of the standard resulted in identifying five areas of impact in two parts: the hazard analysis from the concept phase (Part 3) and the software development phase (Part 6). We describe the five areas of impact with corresponding recommendations in the following subsections. Where it is relevant, we also indicate the levels of autonomy (i.e., SAE J3016 [19] levels 0-5) to which the impact applies.

Identifying Hazards

ISO 26262 defines a hazard as "a potential source of harm caused by malfunctioning behaviour of the item where harm is physical injury or damage to the health of persons" [3, Part 1]. The use of ML can create new types of hazards. One type of such hazard applicable at automation levels 1-3 is caused by the human operator becoming complacent because they think the automated driver assistance (often using ML) is smarter than it actually is [20]. For example, the driver stops monitoring steering in an automated steering function. On one level, this can be viewed as a case of "reasonably forseeable misuse" by the operator, and such misuse is identified in ISO 26262 as requiring mitigation [3, Part 3]. However, this approach may be too simplistic. As ML creates opportunities for increasingly sophisticated driver assistance, the role of the human operator becomes increasingly critical to correct for malfunctions. But increasing automation can create behavioural changes in the operator, reducing their skill level and limiting their ability to respond when needed [21]. Such behavioural impacts can negatively impact safety even though there is no system malfunction or misuse.

Other new types of hazards are due to the unique ways an ML component can fail at higher automation levels (i.e., 4 and 5). For RL, faults in the reward function can cause surprising failures. An RL-based component may negatively affect the environment in order to achieve its goal [5]. For example, an ADS may break laws in order to reach a destination faster. Another possibility is that the RL component *games* the reward function [5]. For example, the ADS figures out that it can avoid getting penalized for driving too close to other cars by exploiting certain sensor vulnerabilities so that it cannot "see" how close it is getting. Although hazards such as these may be unique to ML components, they can be traced to faults, and thus they fit within the existing guidelines of ISO 26262.

Recommendations for ISO 26262: The definition of hazard should be broadened to include harm potentially caused by complex behavioural interactions between humans and the vehicle that are not due to a system malfunction. The standard itself takes note that the current definition is "restricted to the scope of ISO 26262; a more general definition is potential source of harm"[3, Part 1]. The definition and methods for identifying such hazards should be informed by the research specifically on behavioural impacts of ADAS [22] as well as human-robot interaction (HRI) [23] more broadly. For example, van den Brule et al. [24] study how a robot's behavioural style can affect the trust of humans interacting with it.

Faults and Failure Modes

ISO 26262 mandates the use of analyses such as Fault Mode Effects Analysis (FMEA) to identify how faults lead to failures that may cause harm (i.e., are hazards). We can ask whether there are types of faults and failures that are unique to ML and not found in programmed software. Specific fault types and failure modes have been catalogued for NNs (e.g., [13, 15]). Some of these are just "apparent" ML specific faults. For example, a neuron that randomly changes its connection in an operational NN is not really about neurons but rather a conventional fault that can occur in the software on which the NN runs. Others are distinctly ML-specific such as faults in the network topology and learning method that lead to poor generalization (e.g., insufficient connectivity between layers, too high a learning rate, etc.) or faults in the training set. These include inadequate representativeness of the operational environment by the training set, insufficient coverage of rare cases and lack of handling for distributional shift.

Although ML faults have some unique characteristics, this cannot be said about failure modes. All that ML faults can do is to increase the error rate of the deployed component, and thus cause one particular type of failure - an incorrect output for some input. But since most software failures take the form of incorrect output for a given input, we may conclude that there is nothing different about the failure analysis of an ML component as compared to a programmed component, and existing ISO 26262 recommendations apply.

Recommendations for ISO 26262: The distinctive types of ML faults create the opportunity to develop focused tools and techniques to help find faults independently of the domain for which the ML model is being trained. For example, Chakarov et al. [25] describe a technique for debugging misclassifications due to bad training in data, while Nushi et al. [26] propose an approach for troubleshooting faults due to complex interactions between linked ML components. As these techniques mature, ISO 26262 should be amended to require the use of such techniques for ML components.

When the functionality is complex and the ASIL is high (e.g., at higher automation levels), it is unlikely that the error rate can ever be brought to an acceptably low level only through increasing or improving the training set due to the "curse of dimensionality". Specialized architectural techniques should be required to help mitigate the effects of ML faults and failures. Ensemble methods [27] such as bagging [28] and boosting [29] are mature "fault tolerance" techniques used with ML classifiers that reduce the error rate by fusing the result of multiple weaker classifiers to produce a stronger classifier. The simplex architecture [30] that uses a conservative but verifiably safe controller as a fall-back from a more advanced but unverifiable controller has been proposed as a way to make ML components safe [31]. Other similar "safety envelope" approaches have also been proposed for reinforcement learning. [32, 33].

Specification and Verification

Spanfelner et al. [1] point out that there is an assumption in ISO 26262, given by the left side of the V model (Figure 1), that component behaviour is fully specified and each refinement can be verified with respect to its specification. Note that this assumption is also made in other safety-critical domains such as aerospace [34]. This is important to ensure that a safety argument can trace the behaviour of the implementation to its design, safety requirements and ultimately, to the hazards that are mitigated.

This assumption is violated when a training set is used in place of a specification since such a set is necessarily incomplete, and it is not clear how to create assurance that the corresponding hazards are always mitigated. Thus, an ML component violates the assumption. Furthermore, the training process is not a verification process since the trained model will be "correct by construction" with respect to the training set, up to the limits of the model and the learning algorithm.

A deeper issue, discussed by Spanfelner et al. [1], is that many kinds of advanced functionality needed for higher automation levels require perception of the environment, and this functionality may be *inherently unspecifiable*. For example, what is the specification for recognizing a pedestrian? We might observe that since a vehicle must move around in a human world, advanced functionality must involve perception of *human categories* (e.g., pedestrians). There is evidence that such categories can only partially be specified using rules (e.g., necessary and sufficient conditions) and also need examples [35]. This suggests that ML-based approaches are necessary for implementing this type of functionality.

Recommendations for ISO 26262: The approach required for high ASIL component implementation should be based on the specifiability of the functionality being implemented. For functionality that is fully specifiable, programming must be required. For functionality that admits no complete specification (e.g., perception), ML-based approaches should be allowable, and the complete specification requirement must be relaxed. However, even here, the use of specification in a more limited capacity should still be required where possible. The higher-level safety requirements for an ML component allocated by the architectural level (i.e., before refinement at the unit level) can be specified with completeness and traced to hazards. For example, a component may have the requirement "identify pedestrians" in order that the ADS could avoid harming them.

At the component unit design level, the requirements on sample data can be specified in order to ensure that appropriate training, validation and testing sets are obtained. Subsequently, the data gathered can be verified with respect to this specification. Techniques from black-box software testing such as input domain partitioning may be helpful here [36]. For example, the input domain of a pedestrian classifier could be partitioned by age category, pose (e.g., standing, leaning, etc.), clothing type, etc. The data requirements can specify the relative numbers of samples that should come from each partition to ensure coverage and representativeness of the sample data.

Finally, although complete behavioral specifications are not possible, partial specifications may still be. For example, if a pedestrian must be less than 9 feet tall, then this property can be used to filter out false positives. Such properties can be incorporated into the training process or checked on models after training (e.g., [11]). Some of these recommendations may be addressed in a forth-coming OMG standard relating to sensor and perception issues [37].

Level of ML Usage

Figure 1 identifies an architectural level and a unit (i.e., component) level of implementation. ISO 26262 defines a software architecture as consisting of components and their interactions in a hierarchical structure [3, Part 6]. This component decomposition is important for safety because it allows for easier comprehension of a complex system by human assessors and it permits the use of compositional formal analysis techniques.

ML could be used to implement an entire software system, including its architecture, using an *end-to-end* approach. For example, Bojarski et al. [38] train a DNN to make the appropriate steering commands directly from raw sensor data, sidestepping typical ADS architectural components such as lane detection, path planning, etc.

Here, we may assume that the unit level, in the conventional sense of a distinct component that can be developed independently of the architecture, *no longer exists*. This is the case, even if it is possible to extract and interpret the structure of the trained model as consisting of units with distinct functions, since this structure is emergent in the training process and unstable. If the model is re-trained with a slightly different training set, this structure can change arbitrarily. Note that a DNN does have an architecture in a different sense - the set of layers and their connections. However, since it is the training that actually "implements" the required functionality, this architecture is more of an generic execution layer. Thus, an end-to-end approach deeply challenges the assumptions underlying ISO 26262.

Another challenge with an end-to-end approach is that, in some cases, the size of the training set needs to be exponentially larger than when a programmed architecture is used [39]. This puts additional strain on the already challenging problem of obtaining an adequate training set for safety- critical contexts.

Finally, note that issues with an end-to-end approach can also apply when ML is used at the component level, if components are too complex. For example, at one extreme, the architecture can consist of a single component. ISO 26262 specifically guards against this pitfall by mandating the use of modularity principles such as restricting the size of components and maximizing the cohesion within a component. However, the lack of transparency of ML components can hamper the ability to assess component complexity and therefore, to apply these principles. Fortunately, improving ML transparency is an active research area (e.g., [9, 6]).

Recommendations for ISO 26262: Although using an end-toend approach has shown some recent successes with autonomous driving (e.g., [38]), it is incompatible with the assumptions about stable hierarchical architectures of components. This limits the use of most techniques for system safety and we therefore recommend that ISO 26262 only allow the use of ML at the component level.

Required Software Techniques

Part 6 of ISO 26262 deals with product development at the software level and specifies 75 software development techniques, such as shown in Figure 2, that are used in various phases of the development process in the V model (Figure 1). Of these, 34 apply at the unit level, and the remaining at the architectural level. We performed an assessment of the software techniques to determine their applicability to ML components*. Based on our recommendation above on the level of ML usage, we assumed that ML was only used at the unit level and programming is used at the architecture level to connect components.

The charts in Figure 3 show the results of the assessment for the techniques dealing with the unit level. We classified each technique into one of three categories based on the level of applicability to ML. Category Ok means the technique is directly applicable without modification. Most of these cases are due to the fact that they are black box techniques (e.g., *analysis of boundary values, error guessing*, etc.) and thus, the method of component implementation is irrelevant. However, some white box techniques such as *fault injection* also apply. For example, faults can be injected into an NN by breaking links or randomly changing weights (e.g., [40]). Category Adapt says that the technique can be used for an ML component if it is adapted in some way. For example, the technique *walk-through* cannot be used directly with an NN due to the non-transparency charac-teristic. Finally, category N/A indicates that the technique is fundamentally code-oriented and does not apply to an ML component. For example, *no multiple use of variable names* is meaningful for a program but has no corresponding notion in an ML model.

The results in Chart (a) are grouped by the degree to which the techniques are recom-mended. Recall from background section that each technique is marked as highly recom-mended (++), recommended (+) or no recommendation (o) depending on the ASIL level. The bars in each category show the percentage of techniques that apply when considering all techniques (0,+,++), only the recommended techniques (+, ++), and only the highly recommended techniques (++). Since the degree of recommendation varies by ASIL, each percentage is an average value over all four ASILs with the standard deviation in paren-theses. Note that the standard deviation is 0 for the "all" group since every technique is present for each ASIL. Because of the high standard deviation for the highly recommended group, we have included Chart (b) which gives the actual data for each ASIL in this group.

Chart (a) shows that a significant part of the standard is still directly applicable (category Ok) and there is an emphasis on highly recommended techniques. However, the standard deviation is high and Chart (b) shows that most of these highly recom-mended techniques apply to the lower ASIL values - i.e. they are less relevant from a safety critical perspective. Chart (a) also shows that about 40% of the techniques do not apply at all (category N/A) regardless of the degree of recommendation. In general,

---

* The data is available at https://github.com/rsalay/safetyml

FIGURE 3 Percentage of unit-level software techniques applicable to ML components: (a) values averaged across the four ASILs with standard deviation shown in parentheses; (b) values for each ASIL when only highly recommended techniques are considered.

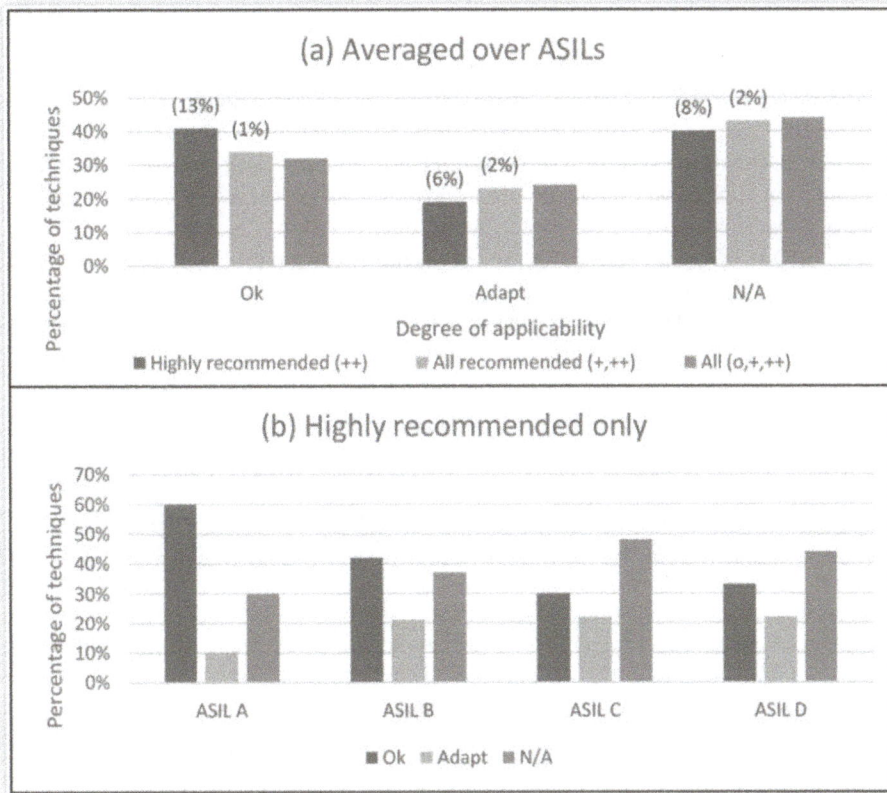

techniques in the software part of the standard are clearly biased toward imperative programming languages (e.g., C, Java, etc.) [34]. In addition to precluding ML components, this bias makes it difficult to accept implementations in other mature programming paradigms such as functional programming, logic programming, etc.

Recommendations for ISO 26262: One approach to addressing the gap in applicable techniques as well as the imperative language bias without compromising safety may be to specify the requirements for techniques based on their *intent* and maturity rather than on their specific details. For example, the intent of the *no multiple use of variable names* technique is to reduce the possibility for confusion that may prevent the detection of bugs. This helps humans understand the implementation better and increase their confidence in its correctness and safety. Thus, the standard can require the use of "accepted clarity increasing" techniques instead of the specific techniques.

Summary and Conclusion

Machine learning is increasingly seen as an effective software implementation technique for delivering advanced functionality; however, how to assure safety when ML is used in safety critical systems is still an open question. The ISO 26262 standard for functional safety of road vehicles provides a comprehensive set of requirements for assuring safety but does not address the unique characteristics of ML-based software. In this paper, we make a step towards addressing this gap by analyzing the places where ML can impact the standard and providing recommendations on how to accommodate this impact. Our results and recommendations are summarized as follows.

Identifying hazards. The use of ML can create new types of hazards that are not due to the malfunctioning of a component. In particular, the complex behavioural interactions possible between humans and advanced functionality implemented by ML can create hazardous situations that should be mitigated within the system design. We recommend that ISO 26262 expands their definition of hazard to address these kinds of situations.

Fault and failure modes. ML components have a development lifecycle that is different from other types of software. Analyzing the stages in the lifecycle reveals distinct types of faults they may have. We recommend that ISO 26262 be extended to explicitly address the ML lifecycle and require the use of fault detection tools and techniques that are customized to this lifecycle.

Specification and verification. Because ML components are trained from inherently incomplete data sets, they violate the assumption in V model-based processes that component functionality must be fully specified and that refinements are verifiable. Furthermore, it is possible that certain types of advanced functionality (e.g., requiring perception) for which ML is well suited are unspecifiable in principle. As a result, ML components are designed with the knowledge that they have an error rate and that they will periodically fail. Rather than disqualifying this class of functionality, we recommend that ISO 26262 provide different safety requirements depending on whether the functionality is specifiable.

The level of ML usage. ML could be used broadly at the architectural level with a system by using an end-to-end approach or remain limited to use at the component level. The end-to-end approach challenges the assumption that a complex system is modeled as a stable hierarchical decomposition of components each with their own function. This limits the use of most techniques for system safety and we therefore recommend that ISO 26262 only allow the use of ML at the component level.

Required software techniques. ISO 26262 mandates the use of many specific techniques for various stages of the software development lifecycle. Our analysis shows that while some of these remain applicable to ML components and others could readily be adapted, many remain that are specifically biased toward the assumption that code is implemented using an imperative programming language. In order to remove this bias, we recommend that the requirements be expressed in terms of the intent and maturity of the techniques rather than their specific details.

Acknowledgment

The authors would like to thank Atrisha Sarkar, Michael Smart, Michal Antkiewicz, Marsha Chechik, Sahar Kokaly and Ramy Shahin for their insightful comments.

References

1. Spanfelner, B., Richter, D., Ebel, S., Wilhelm, U., Branz, W., and Patz, C., "Challenges in Applying the ISO 26262 for Driver Assistance Systems," *Tagung Fahrerassistenz, München* 15, no. 16 (2012).

2. Koopman, P. and Wagner, M., "Challenges in Autonomous Vehicle Testing and Validation," *SAE Int. J. Transport. Safety* 4, no. 1 (2016): 15-24, doi:10.4271/2016-01-0128.

3. International Organization for Standardization, *ISO 26262: Road Vehicles - Functional Safety* (2011).

4. Varshney, K.R., "Engineering Safety in Machine Learning," arXiv preprint arXiv:1601.04126, 2016.

5. Amodei, D., Olah, C., Steinhardt, J., Christiano, P., Schulman, J., and Mané, D., "Concrete Problems in AI Safety," arXiv preprint arXiv:1606.06565, 2016.

6. Tickle, A.B., Andrews, R., Golea, M., and Diederich, J., "The Truth Will Come to Light: Directions and Challenges in Extracting the Knowledge Embedded within Trained Artificial Neural Networks," *IEEE Transactions on Neural Networks* 9, no. 6 (1998): 1057-1068, doi:10.1109/72.728352.

7. Peterson, G.E., Foundation for Neural Network Verification and Validation, *Optical Engineering and Photonics in Aerospace Sensing* (International Society for Optics and Photonics, 1993), 196-207.

8. Rodvold, D.M., "A Software Development Process Model for Artificial Neural Networks in Critical Applications," *International Joint Conference on Neural Networks* 5 (1999): 3317-3322, doi:10.1109/IJCNN.1999.836192.

9. Hendricks, L.A., Akata, Z., Rohrbach, M., Donahue, J., Schiele, B., and Darrell, T., "Generating Visual Explanations," *European Conference on Computer Vision*, Spring, 2016, 3-19, doi:10.1007/978-3-319-46493-0.

10. Huang, X., Kwiatkowska, M., Wang, S., and Wu, M., "Safety Verification of Deep Neural Networks," arXiv preprint arXiv:1610.06940, 2016.

11. Katz, G., Barrett, C., Dill, D., Julian, K., and Kochenderfer, M., "Reluplex: An Efficient SMT Solver for Verifying Deep Neural Networks," arXiv preprint arXiv:1702.01135, 2017.

12. Schumann, J., Gupta, P., and Liu, Y., "Application of Neural Networks in High Assurance Systems: A Survey," *Applications of Neural Networks in High Assurance Systems*, Spring, 1-19, 2010, doi:10.1007/978-3-642-10690-3.

13. Pullum, L.L., Taylor, B.J., and Darrah, M.A., *Guidance for the Verification and Validation of Neural Networks* (John Wiley & Sons, 2007), doi:10.1002/9781119134671.

14. Bedford, D., Morgan, G., and Austin, J., "Requirements for a Standard Certifying the Use of Artificial Neural Networks in Safety Critical Applications," *Proceedings of the International Conference on Artificial Neural Networks*, 1996.

15. Kurd, Z., Kelly, T., and Austin, J., "Developing Artificial Neural Networks for Safety Critical Systems," *Neural Computing and Applications* 16, no. 1 (2007): 11-19, doi:10.1007/s00521-006-0039-9.

16. Martin, H., Tschabuschnig, K., Bridal, O., and Watzenig, D., "Functional Safety of Automated Driving Systems: Does ISO 26262 Meet the Challenges?" *Automated Driving*: 387-416, Spring, 2017, doi:10.1007/978-3-319-31895-0_16.

17. Burton, S., Gauerhof, L., and Heinzemann, C., "Making the Case for Safety of Machine Learning in Highly Automated Driving," *International Conference on Computer Safety, Reliability, and Security*, Spring, 5-16, 2017, doi:10.1007/978-3-319-66284-8_1.

18. Henzel, M., Winner, H., and Lattke, B., "Herausforderungen in der Absicherung von Fahrerassistenzsystemen bei der Benutzung maschinell gelernter und lernenden Algorithmen," *Proceedings of 11th Workshop Fahrerassistenzsysteme Und Automatisiertes Fahren (FAS)*, 2017, 136-148.

19. SAE International, "SAE J3016: Taxonomy and Definitions for Terms Related to on-Road Motor Vehicle Automated Driving Systems," 2017.

CHAPTER 2

20. Parasuraman, R. and Riley, V., "Humans and Automation: Use, Misuse, Disuse, Abuse," *Human Factors: The Journal of the Human Factors and Ergonomics Society* 39, no. 2 (1997): 230-253, doi:10.1518/001872097778543886.

21. Brookhuis, K.A., De Waard, D., and Janssen, W.H., "Behavioural Impacts of Advanced Driver Assistance Systems-An Overview," *EJTIR* 1 no. 3 (2001): 245-253.

22. Sullivan, J.M., Flannagan, Pradhan, A.K. and Bao, S., *Literature Review of Behavioral Adaptations to Advanced Driver Assistance Systems* (AAA Foundation for Traffic Safety, 2016).

23. Goodrich, M.A. and Schultz, A.C., "Human-Robot Interaction: A Survey," *Foundations and Trends in Human-Computer Interaction* 1 no. 3 (2007): 203-275, doi:10.1561/1100000005.

24. van den Brule, R., Dotsch, R., Bijlstra, G., Wigboldus, D. H., and Haselager, P., "Do Robot Performance and Behavioral Style Affect Human Trust?" *International Journal of Social Robotics*, 6, no. 4 (2014): 519-531, doi:10.1007/s12369-014-0231-5.

25. Chakarov, A., Nori, A., Rajamani, S., Sen, S., and Vijaykeerthy, D. "Debugging Machine Learning Tasks," arXiv preprint arXiv 1603:07292, 2016.

26. Nushi, B., Kamar, E., Horvitz, E., and Kossmann, D., "On Human Intellect and Machine Failures: Troubleshooting Integrative Machine Learning Systems," arXiv preprint arXiv 1611:08309, 2016.

27. Ponti Jr, M.P., "Combining Classifiers: From the Creation of Ensembles to the Decision Fusi," *24th SIBGRAPI Conference on Graphics, Patterns and Images Tutorials (SIBGRAPI-T)*, IEEE, 2011, 1-10, doi:10.1109/SIBGRAPIT.2011.9.

28. Breiman, L., "Bagging Predictors," *Machine Learning* 24, no. 2 (1996): 123-140, doi:10.1007/BF00058655.

29. Freund, Y., Schapire, R.E. et al., "Experiments with a New Boosting Algorithm," *ICML, 96*, 1996, 148-156.

30. Sha, L., "Using Simplicity to Control Complexity," *IEEE Software* 18, no. 4 (2001): 20-28.

31. Phan, D., Yang, J., Clark, M., Grosu, R. et al., "A Component-Based Simplex Architecture for High-Assurance Cyber-Physical Systems," arXiv preprint arXiv 1704:04759, 2017.

32. Perkins, T.J. and Barto, A.G., "Lyapunov Design for Safe Reinforcement Learning," *Journal of Machine Learning Research* 3 (2002): 803-832.

33. Fisac, J.F., Akametalu, A.K., Zeilinger, M.N., Kaynama, S. et al., "A General Safety Framework for Learning-Based Control in Uncertain Robotic Systems," arXiv preprint arXiv 1705:01292, 2017.

34. Bhattacharyya, S., Cofer, D., Musliner, D., Mueller, J., and Engstrom, E., "Certification Considerations for Adaptive Systems," *Unmanned Aircraft Systems (ICUAS), 2015 International Conference on*, IEEE, 2015, 270-279, doi:10.1109/ICUAS.2015.7152300.

35. Rouder, J.N. and Ratcliff, R., "Comparing Exemplar and Rule-Based Theories of Categorization," *Current Directions in Psychological Science* 15, no. 1 (2006): 9-13, doi: 10.1111/j.0963-7214.2006.00397.x.

36. Ammann, P. and Offutt, J., *Introduction to Software Testing* (Cambridge University Press, 2016), doi:10.1017/CBO9780511809163.

37. International Organization for Standardization, "ISO/AWI PAS 21448: Road Vehicles - Safety of the Intended Functionality" (under development).

38. Bojarski, M., Del Testa, D., Dworakowski, D., Firner, B. et al., "End to End Learning for Self-Driving Cars," arXiv preprint arXiv 1604:07316, 2016.

39. Shalev-Shwartz, S. and Shashua, A., "On the Sample Complexity of End-to-End Training vs. Semantic Abstraction Training," arXiv preprint arXiv 1604:06915, 2016.

40. Takanami, I., Sato, M., and Yang, Y.P., "A Fault-Value Injection Approach for Multiple-Weight-Fault Tolerance of MNNs," *Proceedings of the IEEE-INNS-ENNS International Joint Conference on Neural Networks* 3 (2000): 515-520, doi:10.1109/IJCNN.2000.861360.

CHAPTER 2

The Development of Safety Cases for an Autonomous Vehicle: A Comparative Study on Different Methods

Junfeng Yang, Michael Ward, and Jahangir Akhtar
Birmingham City Univ.

The Connected and Autonomous Vehicles (CAVs) promise huge economic, social and environmental benefits. The autonomous vehicles supposed to be safer than human drivers. However, the advanced systems and complex levels of automation could also bring accidents by tiny faults of hardware or errors of software. To achieve complete safety, a safety case providing guidance on the identification and classification of hazardous events, and the minimization of these risks needs to be developed throughout the entire development lifecycle process of CAVs. A comprehensible and valid safety case has to employ appropriate safety approaches complying with the automotive functional safety requirements in ISO 26262. The technical focus of present work is on the comparative study of different safety approaches, in particular, Failure Mode and Effects Analysis (FMEA) method and Goal Structuring Notation (GSN) method that have been employed to generate lists of hazardous events, safety goals and functional safety requirements at the vehicle level. A case study on the safety case development of INISIGHT autonomous vehicle has been carried out using the aforementioned methods. This case study covers the safety argument of battery and charging system that supply the whole electric power for INSIGHT vehicle. The safety of this systems has been assessed along with their potential for malfunction together with the layers of protection. The results and conclusions from case study analyses suggest the safety case of CAVs can be developed in a highly effective manner by employing a combined method of GSN and FMEA.

CITATION: Yang, J., Ward, M., and Akhtar, J., "The Development of Safety Cases for an Autonomous Vehicle: A Comparative Study on Different Methods," SAE Technical Paper 2017-01-2010, 2017, doi:10.4271/2017-01-2010.

Introduction

Rapid growth in personal transport is frightening in terms of the spiraling number of injuries and deaths, global pollution and climate change. Back in 2009, 5.5 million accidents in the USA, involving 9.5 million vehicles, killed ~34k people and injured >2.2M others, including 240k hospital admissions [1]. In addition, cars and trucks are estimated to cause 20% all U.S. CO_2 emissions [2]. For the exploding numbers of cars in the developing world, the statistics are even more terrifying. The CAVs equipped with more sensors to detect other road uses and pedestrians, and much higher levels of computer control promise huge reductions in accidents, congestion, and pollution. For example Google claim their driverless car could reduce accidents by 90%, wasted time and fuel by 90%, and massively increase the utilization of cars, meaning fewer cars overall [3]. With a huge market worth in view, every major car manufacturer in the world is developing CAVs. One estimate for sales of autonomous vehicles is 95 million per year by 2035 [4]. The IEEE predicts that autonomous vehicles could be as much as 75% of the market by 2040 [5]. The automotive industry makes a substantial contribution (>£60Bn) to the UK economy, and is expected to see considerable growth in the next decade and more.

One of the earliest reports on autonomous vehicle appeared on 1948 which concerned the development of cruise control in vehicles [6, 7]. Since then the work has been developed by many researchers to include areas such as mechanical antilock braking, electrical stability control, laser based cruise control, pre-crash mitigation. Some of the first autonomous car projects in the 1980s were the Navlab (1980) and the ALV (Autonomous Land Vehicle) in 1984 that were organized by Carnegie Mellon University (CMU). They have continued to develop the autonomous car since then. Recently the CAVs industry keeps blooming and many companies including Mercedes-Benz, General Motors, Continental Automotive Systems, Autoliv Inc., Bosch, Nissan, Toyota, Audi, Volvo, Google and Tesla have developed autonomous cars [8, 9].

Figures for UK's CAVs development show a similar trends. Within UK, the Centre for Connected & Autonomous Vehicles, CCAV, has been established to help ensure UK's world leadership in developing and testing connected and autonomous vehicles. Since 2015, CCAV has continuously funded a series of projects, e.g. GATEway, Venturer, UK Autodrive, INSIGHT, i-MOTORS and FLOURISH, on intelligent mobility research and development. Among them, the INSIGHT project aiming to develop driverless shuttles with a particular focus on improving urban accessibility for disabled and visually-impaired people will be thoroughly introduced. And the safety case developed for the INSIGHT pod will be discussed as the case study in the following sections.

The INSIGHT [10] project is a collaborative project to develop existing autonomous vehicles for safe, slow speed (max 15 MPH) operation in pedestrian areas and pavements, with connectivity not only to control and manage the vehicles, but also for innovative data collection and presentation applications that interact with users and other customers of the systems. An existing electric connected & autonomous vehicle design [11] has been upgraded with advanced sensors to detect and recognize pedestrians, cyclists, mobility scooters, and other vehicles on adjacent roads. These detection capabilities will enable more advanced decision making and a more nuanced approach to way finding, and a smoother ride rather than the simple start/stop common in such systems today. The general view of INSIGHT autonomous vehicle can be seen in Figure. 1.

The INSIGHT pod vehicle is driverless and self-steering (autonomous driving SAE Level 5 [12]), electrically-powered light-weight vehicle designed to carry up to four people and their luggage (including pushchairs and bulky items), see Figure 2 for the general interior view. While the INSIGHT pod is suitable for almost any age group, it has been designed with inclusive at its heart. The vehicle has wheelchair access and, INSIGHT will look specifically at its use by the elderly and those who need assistance in transport, for example the visually impairment. The pod will not just assess the physical passenger experience, such as internal comfort and safety, but also the supply journey information such as calling response times, destination, connections and other support information, all delivered by a human voice interface.

The project activities require a safety case to be made before commencing in order to provide assurance that any reasonable residual risks have been minimized and where possible avoided all together. In addition, a safety case based upon the road vehicle functional safety standards is also required to demonstrate that the vehicle can be safely and reliably driven. The typical approaches documenting safety cases include textual format, tabular form e.g. using FEMA, graphical notations, e.g. GSN. All these approaches have been employed to formulate the safety cases for various autonomous vehicles. For instance, LUTZ pathfinder automated vehicle [13] has developed a defendable safety case using FEMA approach together with a tailored application of ISO26262 automotive functional safety standard [14], and concluded the use of human intervention is required for trials. Another

FIGURE 1 General exterior views of the INSIGHT vehicle.

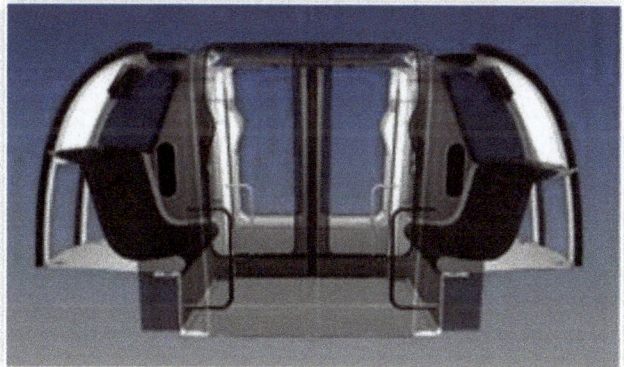

FIGURE 2 General interior views of the INSIGHT vehicle.

similar project, ULTra CAVs providing personal rapid transition between the T5 Business Car Park and Terminal 5 of Heathrow Airport generated a safety case using a combined method of FEMA and GNS. The ULTra CAVs has been safely running since 2011 and delivered in excess of 3.5million passengers, which provide a strong and convincing evidence on that successful safety case.

The INSIGHT vehicle is based upon an existing design ULTra CAVs, where the vehicle supposed to operate in an unconstrained pedestrian area instead of is confined to a well-defined purpose built track. It is necessary therefore that an appropriate safety is developed for INSIGHT vehicle. The present work seeks to explore various approaches for developing a safety case for INSIGHT vehicles. These approaches will incorporate with the general safety management follows a diverse set of legislation and guidance, e.g. SAE J3018 and J3061 [15], UK Code of Practice for Testing Driverless Car [16] and the Road Traffic Act [17]. The performance of various approach will be compared and analyzed.

Vehicle Layout and ISO26262

Vehicle Control System and Propulsion System

In order to allow level 5 autonomous vehicle operation, an autonomous control system has been developed for INSIGHT pod vehicle. This autonomous control system consists

FIGURE 3 Basic functional topology of an INSIGHT vehicle control system.

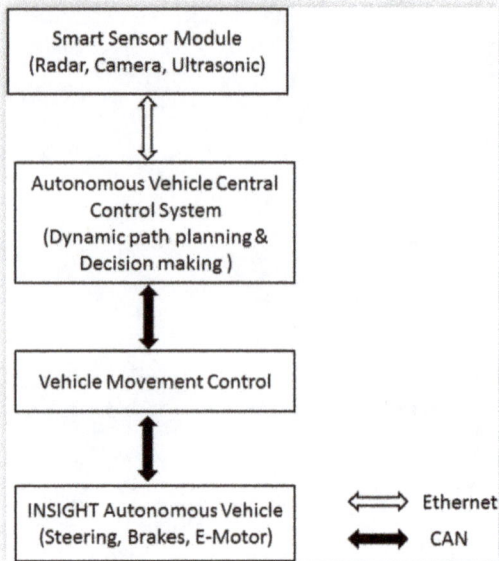

of situational awareness system, central control system facilitating dynamic path planning and decision making, and vehicle movement control system. The smart sensor module ((a group of sensors, e.g. long- and short- range radars, front-, rear-, and side- stereo-cameras and ultrasonic sensors)) connected via wired Ethernet to autonomous vehicle central control system to improve path planning and decision making. The module integrates steering, brakes and E-Motor which respond to demand from the vehicle control system, transmitted via a CAN bus. Figure 3 presents the basic control system architectural of an INSIGHT vehicle. Note that this basic control system is expressed exclusive of the environmental monitoring sensor system, human-machine- interaction control and 4D tactile system for a clearer view.

The INSIGHT vehicle has two Li-ion battery units, high voltage (48V) electric power for traction power system, and low voltage (24V) electric power for vehicle control system and door actuation system. The typical propulsion system for an INSIGHT vehicle is shown in Figure 4. As can be seen, the motor control module converts the 48V DC battery power into low voltage 3 phase AC power while simultaneously controlling motor torque speed and direction. The AC E-Motor drives the vehicle through the front wheels via a fixed ratio transmission and a differential mounted in a transaxle. Note that, the nominal system voltage is restricted to 48 volts to minimize the shock risk.

ISO26262 Road Vehicle Functional Safety Standard

ISO26262 is an adaption of IEC 61508 [18] to meet specific needs of automotive industry. It is the first comprehensive standard that addresses safety related automotive systems comprised of electrical, electronic, and software elements that provide safety related functions. It seeks to address the following important challenges in today's road vehicle

FIGURE 4 Typical propulsion system schematic for an INSIGHT vehicle.

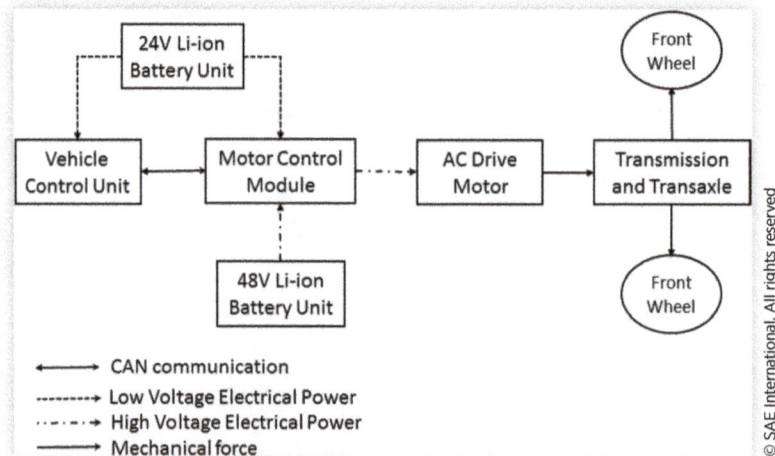

System lifecycle approach to safety: used throughout a safety case.

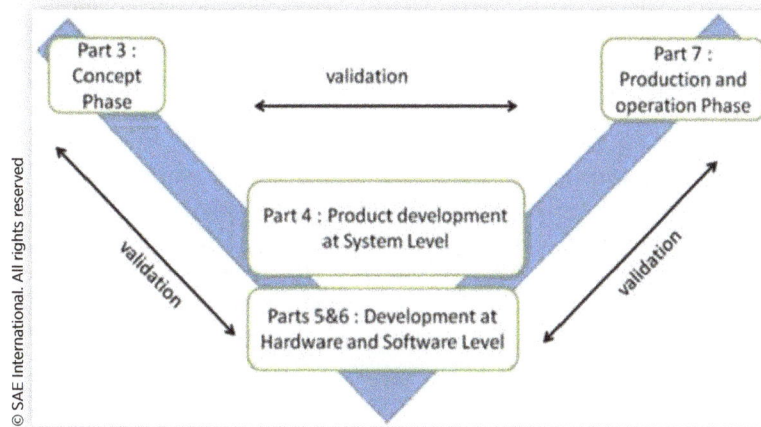

technologies: the safety of new electrical and electronics hardware and software functionality in vehicles; the trend of increasing complexity, software content, and mechatronics implementation; the risk from both systematic failure and random hardware failure. It also provides guidance on how to avoid risk in creating safety-critical systems and regulates critical testing processes.

ISO 26262 defines a safety case as an "argument that the safety requirements for an item are complete and satisfied by evidence compiled from work products of the safety activities. Figure 5 present a system lifecycle approach (V-model) used throughout a safety case. This lifecycle model represents the development of the system from first concept to operation. The concept phase (Part 3) refers to the initial big picture of autonomous vehicle in terms of styling and functionality, etc. Parts 4-6 refers to the vehicle develop and software/ hardware development. Part 7 refers to the final product. Validations refer to various trials, e.g. commissioning test (vehicle shakedown), on-road tests of hardware/software, and trial under a public pedestrian area at different phases. The V-shape is due to the fact that the testing and verification steps are performed in reverse order from design and implementation.

Failure Model and Effects Analysis Method

Failure Mode and Effects Analysis (FEMA) method developed initially for analyzing malfunctions of military systems [19] uses a structured, systematic spreadsheet to documents all the possible failures, risk assessment and management strategies in a design, a manufacturing or assembly process, or a product or service.

The typical FMEA spreadsheet captures all systems/components information including Items, Functional Requirements, Failure Modes, and Causes of Failure. Each possible Cause of Failure has an associated Risk to it which is derived from its Occurrence & Severity. After this the first focus is on Design actions, which means going through the Causes of Failure and mitigate these risks down to the lowest possible level. After the design actions have been considered, the part should be ready for validation through either physical testing or rationale, proving its robustness and ability to meet the safety performance requirements. During Validation, design review, customer reviews/testing issues/concerns may be raised. These issues or failures need to be fed into FMEA ensuring that additional risk is added to the parts of concern proving that we have mitigated that failure and are fit to continue the validation process. The example spreadsheet of FMEA method was given below

Taking the advantages of intuitively clear and high viability, FMEA method has been further developed and adopted by the aerospace and automotive industries.

Goal Structuring Notation Method

Goal Structuring Notation (GSN) method [20] is a graphical notations for the representation of arguments, which was first proposed by T. P. Kelly [21] under the inspiration of Toulmin's argument model [22]. GSN method employs a simple notation of argumentation structures that have been proven to be effective for provides objective safety evidence, therefore has been widely used for developing safety cases for the industrial use and research purpose. Recently, GSN method has been incorporated into ISO26262 to satisfy the critical safety assurance of automotive systems, e.g. start/stop system, and EPS system [23, 24, 25, 26, 27, 28].

Typically, GSN method consists of a group of symbol of notations linked by directional arrows explicitly representing the individual elements (safety goals, solution, context and strategies) of an argument and the relationships between these elements, such as rectangular boxes for safety target, ovals for assumption or justification, circle for evidence (solution), parallelogram-shaped boxes for strategy (argument), rounded-end boxes for context (additional information). An example safety argument constructed using GSN is given below.

As shown in Figure 6, the safety goal of targeted system needs to be claimed by identifying the possible hazards and mitigating them through sufficient and appropriate evidences. Due to the complexity of system, the top-level safety goal usually has to

TABLE 1 Example risk assessment represented using FMEA spreadsheet

Item	Function Requirement	Failure Model	Casual Factor	Immediate Consequences	Severity	Frequency	Exposure	Hazard Ranking	Mitigation
Battery	supply electric power for E-motor	supply insuffcient electric power	over-heating; degradation	vehicle lose power	3	2	2	6	BMS monitory battery voltage and temperature

FIGURE 6 An example safety argument represented using GSN.

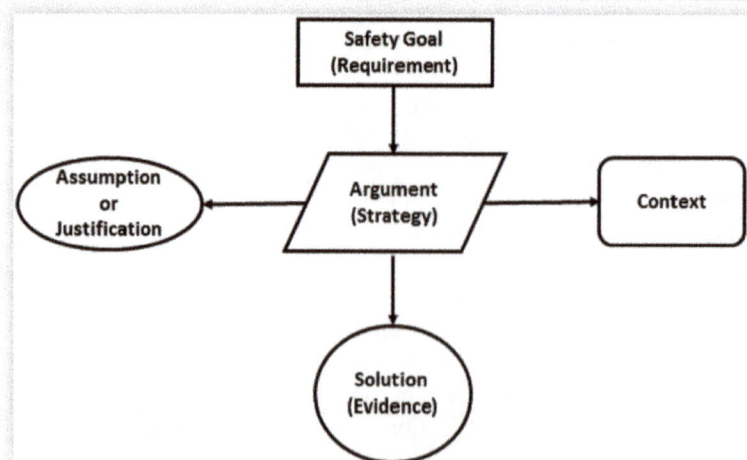

be decomposed into sub-level goals and this decomposition may continue until sub-level claim and evidences asserted.

Both FMEA and GSN have been widely used for the risk assessment of automotive industries, and proven to be valid and capable methods.

The following sections provide a detailed description on a safety case for INISHGT autonomous vehicle constructed using FMEA and GSN methods incorporating with ISO26262 road vehicle functional safety standard. In addition, the performance appraisal of these method provides guidance toward a valid and defendable safety case.

Safety Case Development

In the present work, the safety case investigates and documents all the hazards and risks associated with autonomous vehicle, including mechanical system, electric hardware, application and embedded software, communications, health and safety of passengers, roads users, pedestrians, risk assessments, safety systems of work and insurance and liability.

A valid safety case for an autonomous vehicle consists of four main inter-dependent components, namely:

- Safety target that must be addressed to assure vehicle safety.

- Evidence for the safety target obtained from study, analysis and test of the vehicle system.

- Argument showing how the rationale indicates compliance with the safety target.

- Context identifying the basis for the argument presented.

A set of safety targets for the vehicle commissioning tests is generated with the objective of achieving acceptable safety considering the prototype nature of INSIGHT vehicle. Based on the goals of the safety work, the principles of safety process rationale argument were chosen as:

1. Hazard Generation. Identifying the vehicle operational situation and the possible hazardous events associated with safety targets.

2. Risk Assessment. Classifying each hazard in terms of frequency of occurrence, severity of resulting harm and controllability of hazard, and determining the Automotive Safety Integrity Level (ASIL) of system by considering the SAE J2980 standard [29].

3. Hazard Management. Addressing safety requirements through an appropriate combination of system design in accordance with the ASIL indicated.

An ASIL shall be determined for each hazardous event based on its severity level (S1-S3), probability of exposure (E1-E4) and controllability level (C1-C3) in accordance with Table 2. The number 1 represents lowest level and 4 the highest one. The classification of severity, exposure and controllability are given in SAE J2980.

As can be seen, Four ASILs are defined: ASIL A, ASIL B, ASIL C and ASIL D, where A representing the least stringent level and D the most stringent level. QM indicates quality management system can be sufficient to develop element(s)

TABLE 2 ASIL Determination per ISO26262:2011

ASIL Classification		C1	C2	C3
S1	E1	QM	QM	QM
	E2	QM	QM	QM
	E3	QM	QM	A
	E4	QM	A	B
S2	E1	QM	QM	QM
	E2	QM	QM	A
	E3	QM	A	B
	E4	A	B	C
S3	E1	QM	QM	A
	E2	QM	A	B
	E3	A	B	C
	E4	B	C	D

that implement the safety requirement allocated to the intended functionality. Or it can support the rationale for the independence between the intended functionality and the safety mechanism.

Figure 7 summarizes the safety process rational argument employed in the present work. If a system has high ASIL and subjects to the constraints, its safety requirement can be decomposed by multiple redundant subsystem working together, each with a lower ASIL. This process is so-called ASIL decomposition that allows the best safety strategies to be developed efficiently.

Case Study

The INSIGHT autonomous vehicle has two Li-ion battery units mounted on to the battery tray of the rear. The battery and charging system supply as one of the most important systems supply the whole electric power for the INISIGHT vehicle. These batteries incorporates an on-board Battery Management System (BMS) in the vehicle. The BMS has the feature of measuring cell voltage and temperature, performing cell balancing function and monitoring the cell fault conditions, providing these information to the external systems via CAN. Since the Lithium-ion batteries contain flammable electrolyte and may pose a fire/explosion and other hazards when it's overheated or short-circuited. Thereafter the safety case must include the risk assessment on battery and charging system. This section takes the battery and charging system as the case study to examine the product-based safety rational argument.

Table 3 present the FMEA-based safety argument for the battery and charging system. The potential malfunctioning behavior relevant to the battery and charging system and potential hazards have been identified. Then severity level, controllability and probability of exposure of this vehicle hazard and ASIL level are classified for each hazards. Failure of the battery and charging system is clearly undesirable. However system design features mitigate the risk associated with such failures to acceptable levels. Figure 8 shows GSN-based product argument structure for the battery/charging Systems Safety argument. This valid argument (strategy) is supported by evidence in compliance with ISO 26262.

As can be seen clearly, FMEA is better for documenting the evidence and context. But it seems hardly to convey all necessary information effectively along. While, GSN are better for presenting a decision making process, especially the decomposition of safety goal on a complex system which involves a number of risks. However, for a complex system, the GSN has to cover a number of sub-goals which may cause an intricate GSN structure and make it difficult to follow.

TABLE 3 FMEA-based safety argument for the battery and charging system of INSIGHT vehicle

Entry id	Item	Cause of failure	Consequence	Mitigation	Controllability	Severity	Frequency	ASIL Ranking	General Comments
1	Battery/ charger	Electric current leakage	Electric shock	Nominal system voltage is restricted to a non-lethal level (48volts). No passenger exposure to high voltage. No circuits used within the passenger compartment operate at voltages above 24V. Vehicle charging contacts are mounted under the vehicle and are inaccessible to passengers. Neither vehicle nor the mounted charging contacts are live when they are not connected together	C1	S3	E1	QM	
2	Battery/ charger	Resistive connection	Overheating	Temperature sensing employed to detect excessive temperature at charging contacts. Purpose designed connectors used for all high current connections.	C1	S1	E3	QM	
3	Battery/ charger	Rapid charging or discharging	Fire/explosion	Battery protection fuse mounted within battery pack. The battery pack is physically separated from the passenger compartment by bulkheads.	C1	S3	E3	A	
4	Battery/ charger	Battery heating	Fire/explosion	Battery chargers automatically control charging to avoid overcharging. Vehicle controller continuously monitors individual battery voltage and temperature; it will stop charging if overcharging or over temperature fault conditions are detected.	C1	S3	E2	QM	

CHAPTER 3

FIGURE 8 GSN-based Safety goal rational argument for a battery /charging system of INSIGHT vehicle.

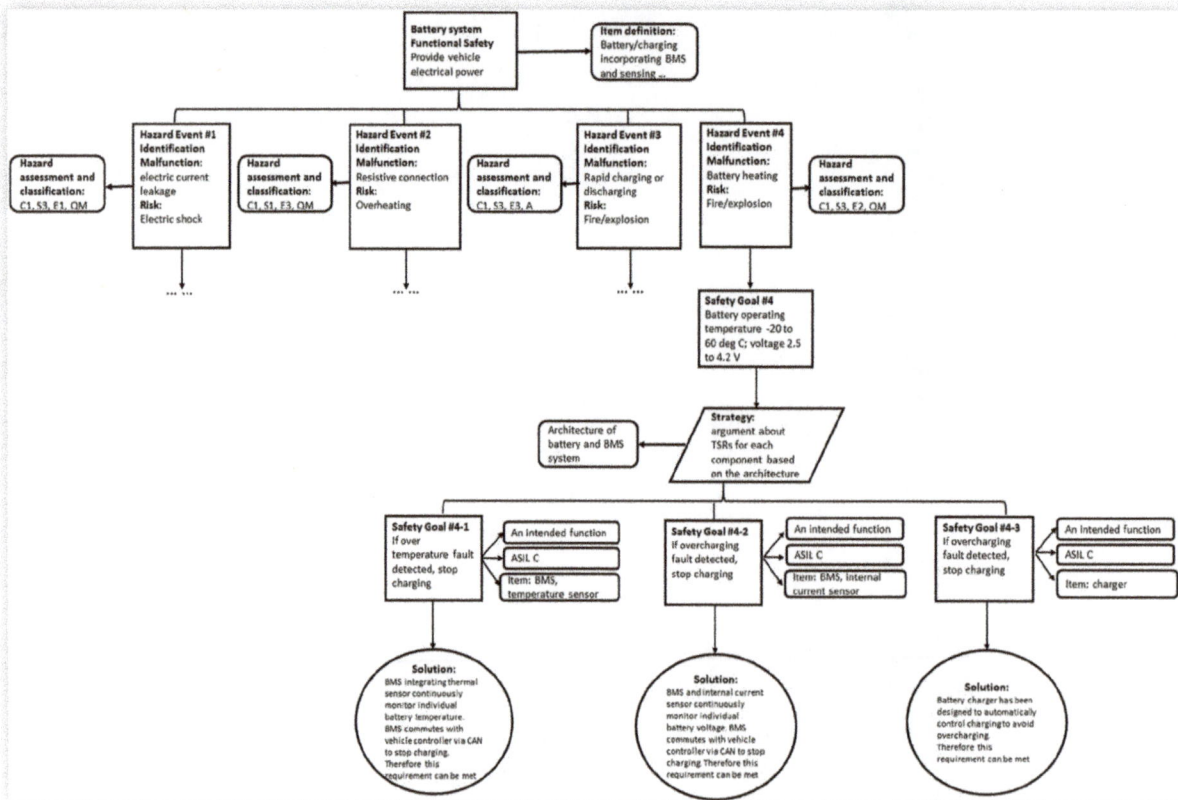

Conclusions

Safety has been a prime consideration throughout the development of the INSIGHT autonomous vehicle. The present work has developed a safety case in order to provide assurance that any reasonable residual risks have been avoided during the vehicle commissioning test. This safety case builds in accordance with the ISO26262:2011 road vehicle functional safety standard. The principle conclusion of safety assessment is that no features of the INSIGHT design concept and operating concept would indicate that the level of safety of the INSIGHT system would be unacceptable. The INSIGHT vehicle is therefore acceptably safe to commence operations.

A diverse approach to the assessment of the safety of the system has been adopted, including assessment against FMEA, GSN and SAE guidance, and quantified risk assessment. The risks associated with all the identified hazards are considered to be in the tolerable or acceptable risk categories. A case study on a battery and charging system of INSIGHT vehicle demonstrates the typical structure of safety goal rational argument. The analysis results indicated that the approach adopted was appropriate.

Regarding the performance appraisal of FMEA and GSN methods, it found FMEA is better for documenting the evidence and context. But it seems hardly to convey all necessary information effectively along. While, GSN are better for presenting a decision making process, especially the decomposition of safety goal on a complex system which involves a number of risks. Hence, a combined method of FMEA and GSN is suggested to contracture a valid and defendable safety case in an efficient and effective manner.

In terms of future work, we would like to continue our safety case to explore the aforementioned method to address the unique aspects of CAVs, e.g. navigation system and decision-making system, which have not been thoroughly discussed.

Contact Information

Junfeng Yang, PhD
School of Engineering and the Built Environment
Faculty of Computing, Engineering and the Built Environment
Birmingham City University
City Centre Campus
Millennium Point
Birmingham B4 7XG
United Kingdom
Phone: +44 (0)121 300 4293
Junfeng.Yang@bcu.ac.uk

Acknowledgments

This project was funded by Innovate UK (grant agreement No. 102583) and supported by the Centre for Connected and Autonomous Vehicles, UK.

Definitions/Abbreviations

CAV - Connected and Autonomous Vehicle
FMEA - Failure Mode and Effects Analysis
GSN - Goal Structuring Notation
ASIL - Automotive Safety Integrity Level
BMS - Battery Management system

References

1. http://www-nrd.nhtsa.dot.gov/Pubs/811363.pdf.

2. http://www.ucsusa.org/our-work/clean-vehicles/car-emissions-and-global-warming.

3. http://www.google.co.uk/about/careers/lifeatgoogle/self---driving-car-test-steve-mahan.html.

4. http://www.navigantresearch.com/research/autonomous---vehicles.

5. http://www.ieee.org/about/news/2012/5september_2_2012.html.

6. "keesler news," keesler, 1 1 1948. [Online], accessed 13 10 2016, http://www.keesler.af.mil/AboutUs/FactSheets/Display/tabid/1009/Article/360538/history-of-keesler-air-force-base.aspx.

7. Ioannou P.A. and Chien C.C., "Autonomous Intelligent Cruise Control," *IEEE Transaction on Vehicle Technology* 42, no. 4 (1993): 657-672.

8. Thorpe C., Hebert M., Kanade T., and Shafer S., "Toward Autonomous Driving: the CMU Navlab," *IEEE*, (1991): 31-41.

9. Thorpe C., Hebert M., Kanade T., and Shafer S., "Vision and Navigation for Carnegie-Mellon Navlab," *IEEE*, 10, no. 3 (1988): 362-373.

10. Centre for Connected and Autonomous Vehicles, UK, accessed January 2016, INSIGHT Project http://insightcav.com.

11. Heathrow Pod. accessed July 2015, http://www.ultraglobalprt.com/wheres-it-used/heathrow-t5/.

12. SAE International Surface Vehicle Information Report, "Guidelines for Safe On-Road Testing of SAE Level 3, 4 and 5 Prototype Automated Driving Systems (ADS)," SAE Standard J3018™, Iss, March 2015.

13. Peters, A., "Safety of the LUTZ Pathfinder Automated Vehicle," *22nd ITS World Congress, Paper number ITS-2427*, Bordeaux, France, October 5-9, 2015.

14. ISO 26262 -Road Vehicles -Functional Safety. Parts 1 to 12.

15. SAE International Surface Vehicle Information Report, "Guidelines for Safe On-Road Testing of SAE Level 3, 4 and 5 Prototype Automated Driving Systems (ADS)," SAE Standard J3018™, Iss, March 2015.

16. UK Department for Transport, "The Pathway for Driverless Car: A Code of Practice for Testing," 2015.

17. UK Government, Road Traffic Act, 1991.

18. IEC 61508 - Functional Safety of Electrical/Electronic/Programmable Electronic Safety-related Systems.

19. United States Department of Defense, "MIL-P-1629 - Procedures for Performing a Failure Mode Effect and Critical Analysis," Department of Defense (US), MIL-P-1629, November 9, 1949.

20. Goal Structuring Notation Working Group, GSN Community Standard Version 1, http://www.goalstructuringnotation.info/, 2011.

21. Toulmin Stephen, E., *The Uses of Argument* (Cambridge University Press, 1958).

22. Kelly, T.P., "Arguing Safety - A Systematic Approach to Safety Case Management," DPhil thesis YCST99-05, Department of Computer Science, University of York, UK, 1998.

23. Kelly, T. and Weaver, R., "The Goal Structuring Notation-A Safety Argument Notation," *Proceedings of the Dependable Systems and Networks 2004 Workshop on Assurance Cases*, July 2004.

24. SAE International Surface Vehicle Recommended Practice, "Cybersecurity Guidebook for Cyber-Physical Vehicle systems," SAE Standard J3061™, Iss, January 2016.

25. Palin, R. and Habli, I., "Assurance of Automotive Safety: A Safety Case Approach," *SAFECOMP*, Vienna, Austria, 2010.

26. Palin, B., Ward, D., Habli, I., and Rivett, R., "ISO 26262 Safety Cases: Compliance and Assurance," *IET Intl. System Safety Conf.*, 2011.

27. Habli, I. et al., "Safety Cases and Their Role in ISO 26262 Functional Safety Assessment," *32nd International Conference on Computer Safety, Reliability, and Security*, Toulouse, France, 2013.

28. Matsuno, Y., "D-Case Ediotor," http://www.il.is.s.u-tokyo.ac.jp/deos/dcase/.

29. SAE International Surface Vehicle Recommended Practice, "Considerations for ISO 26262 ASIL Hazard Classification," SAE Standard J2980, May 2015.

Autonomous Vehicle Sensor Suite Data with Ground Truth Trajectories for Algorithm Development and Evaluation

William Buller
Michigan Technological University

Helen Kourous and Jakob Hoellerbauer
Ford Motor Company

This paper describes a multi-sensor data set, suitable for testing algorithms to detect and track pedestrians and cyclists, with an autonomous vehicle's sensor suite. The data set can be used to evaluate the benefit of fused sensing algorithms, and provides ground truth trajectories of pedestrians, cyclists, and other vehicles for objective evaluation of track accuracy. One of the principal bottlenecks for sensing and perception algorithm development is the ability to evaluate tracking algorithms against ground truth data. By ground truth we mean independent knowledge of the position, size, speed, heading, and class of objects of interest in complex operational environments. Our goal was to execute a data collection campaign at an urban test track in which trajectories of moving objects of interest are measured with auxiliary instrumentation, in conjunction with several autonomous vehicles (AV) with a full sensor suite of radar, lidar, and cameras. Multiple autonomous vehicles collected measurements in a variety of scenarios designed to incorporate real world interactions of vehicles with bicyclists and pedestrians. Trajectory data for a set of bicyclists and pedestrians was collected by separate means. In most cases, the real-time kinetic receivers on the bicyclists and pedestrians achieve RTK (Real Time Kinematic)-fixed, or RTK-float accuracy, resulting in errors on the order of a few centimeters, or a few decimeters, respectively; position

CITATION: Buller, W., Kourous, H., and Hoellerbauer, J., "Autonomous Vehicle Sensor Suite Data with Ground Truth Trajectories for Algorithm Development and Evaluation," SAE Technical Paper 2018-01-0042, 2018, doi:10.4271/2018-01-0042.

accuracy on the instrumented interaction vehicles is on the order of 10 cm. We describe the data collection campaign at the University of Michigan's Mcity Test Facility for connected and automated vehicles, the interaction scenarios and test conditions, and will show some visualizations of the test as well as initial evaluation results. These data will serve as a global-frame, multi sensor/multi actor canonical dataset which can be used for the development and evaluation of extended-object tracking algorithms for autonomous vehicles.

Introduction

The problems of pedestrian detection, recognition, and tracking have attracted great interest for several decades due to their importance in automotive safety, robotics, and surveillance. In recent years, a number of datasets have been collected and made publicly available for research [1, 2, 3]. However, the focus of these datasets has not addressed sensor fusion and evaluation of track accuracy. In particular, the datasets are exclusively optical, and do not support research on radar and lidar. Additionally, the existing datasets have no independent collection of pedestrian locations, so there is no means to assess the accuracy of the track developed by algorithms under test.

To extend the available datasets for research to include lidar and radar, and provide independently collected ground truth for evaluating track errors, measurements with cameras, lidar, and radar, were collected concurrently, along with RTK corrected locations for pedestrians and bicyclists, in realistic traffic scenarios. Inexpensive RTK systems have been shown to provide position estimates ranging from a few centimeters to a few decimeters [4, 5].

The intended use for this dataset is to provide an externally measured verification of the 3D position of tracked objects in an urban setting for verification of tracking and sensor fusion algorithms. Labeling in the sense of 2D image segmentation of the output of the seven onboard AV cameras will be addressed in future work.

Location

The tests were conducted in 2017 on May 23-25 and September 21-22 at the University of Michigan's Mcity Test Facility for Connected and Automated Vehicles in Ann Arbor, Michigan.

This closed test track encompasses a few city blocks with storefronts (that simulate urban occlusion), configurable traffic lights, roundabouts and unstructured intersections, sidewalks and crossings, active pedestrian crosswalks, bicycle lanes, a tunnel, simulated foliage cover, and a short highway / merge section.

The scenarios were conducted primarily in the urban section, at the intersection of State and Main, and the intersection of State and Liberty, shown in Figure 1.

Experimental Design

Collections were designed to incorporate the sensor suite of an autonomous vehicle, operating in an environment populated with other vehicles, bicyclists, and pedestrians,

The University of Michigan Mcity Test Facility for connected and automated vehicle; shown is a subset of the facility which comprises the urban intersections of interest for studying pedestrian interaction.

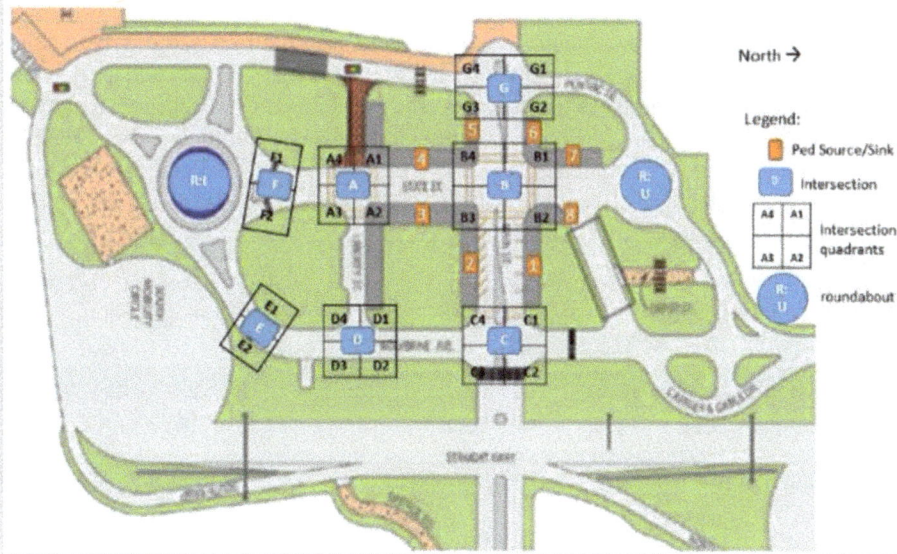

SCENARIO SCHEMA

where the time ordered locations of all the vehicles, bicyclists, and pedestrians, are determined independently. The time-ordered locations are recorded in universal coordinated time (UTC). These independent measurements form the ground truth, by which, estimation algorithms acting on measurements from the sensors can be evaluated.

Scenarios were designed to span a set of test conditions that could be realized in the micro-urban setting at Mcity with a limited number of controlled actors. Specifically we wished to test the following scenarios:

- Sensor range; track-birth range
- Robustness of track segmentation
- Sensor fusion performance
- Challenging maneuvers related to occlusion such as Unprotected Left Turns Across Traffic (LTAT)
- Pedestrian tracking, including clustering/segmentation
- Capturing natural pedestrian behavior
- Bounding box stability under lidar illumination diversity conditions (field of view, illumination angle)
- Sensor range and sensor fusion performance in roundabouts
- Bicycle tracking

The second campaign in September 2017 was focused more on the "organic" naturalistic interactions of pedestrians and vehicles and infrastructure at intersections.

The experiments are organized into Acts, Scenes, and Takes. This serves to assist the execution of the experiments by grouping experiments with the same macro setup as "Acts" (for example, reconfiguring any parked cars or using a different intersection as "scene center").

Scenes are scenarios that begin with a simple set of actions and progress to more complex choreography. This allows actors and drivers to build up competence in the execution of any particular scenario. Finally, each Scene is enacted in two or more "takes" which both allows for any data collection system issues or choreography issues to be worked out, as well as increasing the number of data sets collected on any given scenario, for completeness. A brief description of the test configuration by scenario is given in Table 1.

For uniformity and ease of collection each "take" lasts 5 minutes, the start and end of which is called on a universal radio channel to the nearest UTC second by the Test Coordinator. This procedure is invaluable in the postcollection bookkepping exercise of collating the quantitative log files with the experiment taxonomy. This is done in practice by recording the associateion of the log files from each sensor and the act, Scene, and take in a master spreadsheet termed the "Data Dictionary." This Data Dictionary

TABLE 1 Description of the scenarios, each test's purpose, as well as notes on the test configuration, and the actors involved. AV refers to Autonomous Vehicle with full complement of sensors described above

Scenario	Test	Configuration	Actors
Act I, Scene 1	Sensor range (Vehicles)	Stationary Ego	Two instrumented AVs
Act I, Scene 2	Sensor range (Pedestrians)	Stationary Ego	Five instrumented pedestrians
Act I, Scene 3	Occlusion (Vehicle)	Stationary Ego	Two instrumented AV in LTAT occluded configuration
Act I, Scene 4	Segmentation	Stationary Ego	Two instrumented AV traveling in close proximity
Act II, Scene I	Occlusion, range	Simple ego motion on State; Parked cars	Two instrumented AV interacting with Ego at intersections; Groups of pedestrians cross at A and B
Act II, Scene II		Simple ego motion, parked cars	Addition of bicycles to Scene 1; interact with parked cars
Act II, Scene 3	Range on passing maneuvers	Highway passing	Two AV passing Ego on straightaway
Act III, Scene 1	Adaptive cruise control (ACC)	Ego as follower in congested intersections	Two AV occluding sensors of ego car
Act III, Scene 2	ACC	Ego as follower in congested intersections	Two AV occluding sensors of ego car; presence of parked cars
Act III, Scene 3	Tracking in congested urban scenario	Chaotic pedestrian behavior with "jaywalking"	Two AV cars; 8 pedestrians and two bicycles in driving lanes
Act IV, Scene I	Sensor range, aspect change	Interactions at Roundabouts	All three AVs
Act IV, Scene 2	Sensor range	Interactions at Roundabouts	All three AVs, 8 pedestrians,
Act IV, Scene 3	Sensor range	Interactions at Roundabouts	All three AVs, 8 pedestrians, 2 bikes; jaywalking
Act V, Scene 1	Lidar/Radar fusion	Stationary ego	AVs executing "figure eight" maneuvers in med and far radar range
Act VI, Scene I	Radar field of view/range	Highway interactions	All three AVs, collecting additional radar detection-level data

can also be used to automate the bactch processing and transformation (coordinate frames and time base) of the various data files.

Autonomous Vehicle Sensors

Sensors on the Autonomous Vehicle 2nd generation research fleet consist of an array of seven machine vision grade cameras (Flea3), Four Velodyne HL32 lidar sensors, and two automotive long/medium range scanning radars. The physical arrangement of the sensors and their fields-of-view is shown below in Figure 2; however note that the field-of-view representations are not to scale. Example visualizations using the combined returns from the suite of sensors are shown in Figure 3 and Figure 4.

CAMERAS

The camera data are stored directly in the LSHM (a proprietary shared memory frame-work) native format. Rectified PNG (portable network graphic) files can be generated from native LSHM image data, at effective rate of 25fps (tasked at 30 Hz but there are some losses due to processing and pairing of frames between cameras) for each of the roof rack cameras, and 5.3 Hz for the high resolution center camera (tasked at 6 Hz but less due to the same reasons). The filenames correspond to the times of the trigger signal for each set of camera images.

RADAR

The Ford development autonomous vehicles are equipped with forward- and rear-facing ESR automotive radars, with both long-range (174 m) and medium range (60 m) mode. The radar data are collected at 64 tracks per 50 ms scan in the native Lightweight Communications and Marshalling (LCM) log file, in the local reference frame of the vehicle. LCM is a protocol for message passing and data marshalling for high bandwidth, low latency realtime applications.

FIGURE 2   Diagram of AV Sensor Coverage (range not to scale).

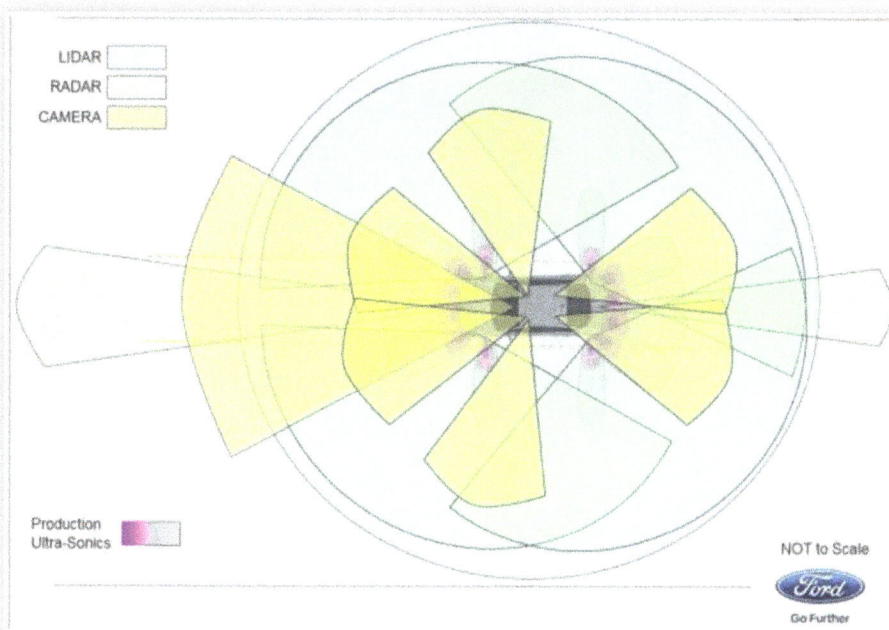

FIGURE 3 Example visualization of AV sensor data - the pedestrian is apparent in the front stereo cameras as well as in lidar and radar (dark blue circle icon in top image near the "feet" of the pedestrian point cloud).

FIGURE 4 Example visualization of AV sensor data from a more complex scenario; there is another AV, pedestrians on the sidewalk, and a bicycle in the lane of travel.

LIDAR

Data were also collected with the four Velodyne HDL32 Lidar units collecting 10 revolutions per second. These data are timestamped in the local unix epoch and projected in the vehicle local frame.

Screen-captured movies of these multisensory visualizations will also be provided in the open source dataset.

Ground Truth Collection - AV Trajectories

Each of the vehicles is equipped with an Applanix GPS/IMU in addition to a sensor suite consisting of seven machine vision cameras, four Velodyne lidars and two automotive radars. For nominal autonomous operation, only the IMU is used to estimate pose, which is then corrected using intensity matching between LiDAR scans and the previously generated prior map. For this exercise, the GPS was used in differential-correction mode to provide vehicle tracks in the global reference frame. The trajectory was continuously updated using a local Continuously Operating Reference Station (CORS) unit 400 m east of Mcity operated by the University of Michigan. It broadcasts RTCM (Radio Technical Commission for Maritime Services) 2.3 and 3.1 corrections in the WGS84 (World Geodetic System) datum using the NTRIP (Networked Transport of RTCTM via Internet Protocol). Thus the intensity based localization can also be corroborated with the high-accuracy differential GPS.

GROUND TRUTH COLLECTION - PEDESTRIAN AND CYCLIST TRACKS

Pedestrians and bicyclists were equipped with Reach RTK units by Emlid. The RTK data can be corrected by a suite of algorithm [6] to achieve accuracy with either integer (a.k.a. fixed), or floating ambiguity resolution. Integer ambiguity resolution provides standard deviations less than 10 cm, while float ambiguity resolution standard deviations are typically 10 to 45 cm [7]. The correction quality that can be achieved depends on atmospheric propagation channel properties, and is impacted by multiple factors, including cloud coverage, visibility of satellites, and multi-path fading in the environment.

Data is recorded for each of the roving units (rovers), along with an additional stationary unit (base station), shown in Figure 5, which serves as the base station for differential correction. This base station's location appears as the Reference Point in Figure 11.

The locations of the RTK units are recorded as ephemeris files in human readable ASCII format, where each row records a time-sample of data. The update rate is typically 5 to 10 Hz.

The columns contain time and position information, along with quality metrics, as follows:

- Time (UTC)

- Position: Latitude (deg), Longitude (deg), Height Relative to Sea Level (m)

- Correction Quality Metrics: Q $Q\in [1,...6]$, where 1:fix, 2:float, 3:SBAS, 4:DGPS, 5:single, 6:PPP

- Number of Satellites received

- Measurement covariance: Standard Deviations in local coordinates, North, East, Up, North-East, East-Up, Up-North, in meters

FIGURE 5 A single RTK unit is used as the base station for differential correction. The antenna is on the copper plate in the foreground.

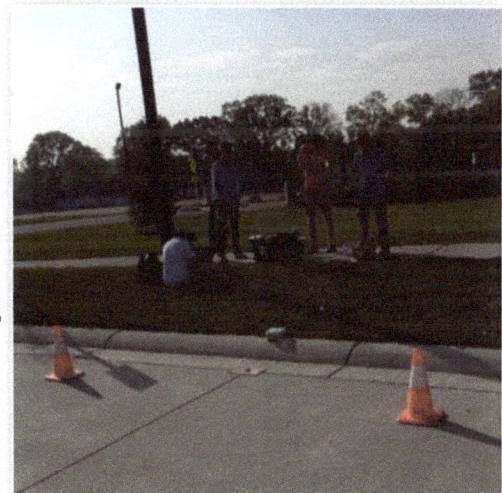

Photo of participating pedestrian holding RTK unit. The RTK unit is attached to a battery pack and an antenna. The battery pack is carried in the participant's pocket. In this case, the antenna is affixed to the top of the participant's hat.

- Time difference between rover and base station data epochs (referred to as age) in seconds;
- Ratio factor used to validate choice of correction method, see equation 7 in [8].

To achieve the best possible accuracy for the RTK units, the antennas are mounted on conducting plates to mitigate ground reflections and improve positioning accuracy. The antennas were placed on top of foil lined clothing, or hats, as in Figure 6. It is important to note that many pedestrian detection and tracking algorithms rely on machine learning via training on exemplar pedestrians; therefore it is important to appear as "typical" as possible (namely without obvious or large instrumentation). The power packs and processors are quite small and fit unobtrusively in pockets or under clothing, while the antennas themselves are fitted onto hats, shoulder pads, backpacks, etc.

Traffic Light Phase Data

For some of the scenarios we were interested in pedestrian decision making based on traffic light state as well as in the presence of vehicles at or near the intersection. Recordings of the traffic light states were stored during the collections in September, and a plot of the data are shown in Figure 7.

Aerial Observation

In addition to the quantitative measurements undertaken by the vehicle sensors and rover units, a subset of the scenes were recorded from a stationary rotor UAV [9] from altitudes between 80 to 120 m. The imagery provides context for the data set, and cross

Millisecond resolution traffic light (and thus pedestrian right of way) states also coordinated to UTC time allow interactions to be studied in the context of infrastructure state.

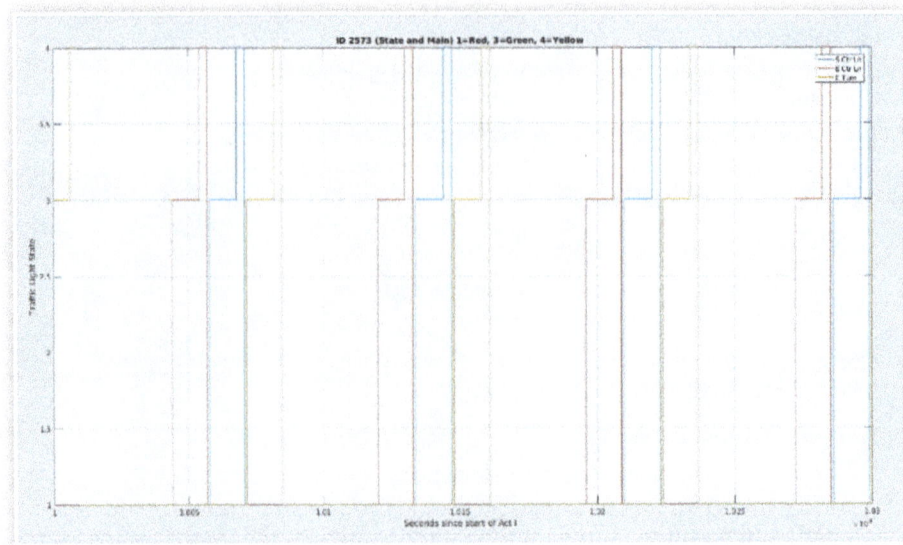

Video frame from UAV showing multiple vehicles and pedestrians during collection at Mcity. The field of view includes the intersections of Main and Liberty with State St.

validation of the vehicle and pedestrian trajectories. A frame from the video with three of the test vehicles, and multiple pedestrians appears in Figure 8. Although it is difficult to visually locate the pedestrians in the static image, they are quite apparent in the videos. The intersections are labelled as A and B in Figure 1.

A Note about Coordinate Frames

Ford AV sensor data is collected in the sensor frame, and is stored in the native LCM message format after transformation (using extrinsic calibration parameters) to a linearized global frame, Figure 9, that is vehicle and mission defined. The public domain version of the dataset described herein will contain sensor data in the common global frame as well as time stamp keys that allow times to be referenced in UTC.

Vehicle sensor data is collected in the sensor frame and stored in the local frame; the global-to-local transform for each measurement is derived from either high-accuracy (differentially corrected) GPS, or a localization algorithm which allows sensor data to be transformed into the common GPS frame.

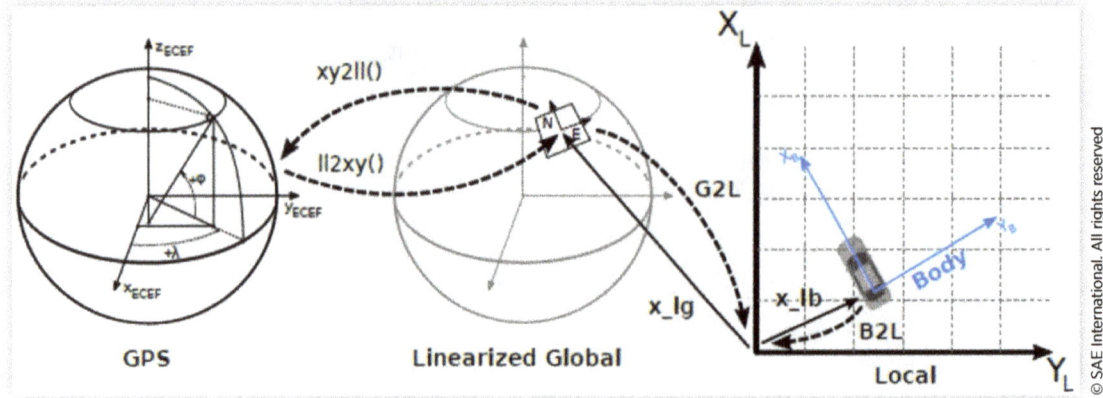

These data were extracted from the LCM log, and were exported at the frequency at which the POSE message is updated (~5 ms or 200 Hz) as a comma-delimited text file. Due to the large size of exported serialize lidar data, we will explore alternative formats such as LAS.

We have extracted the corresponding GPS times as well as Pose of the vehicle in global GPS and LOCAL frame. The vehicle sensor data are recorded in GPS time, while the RTK rovers are recorded in UTC. Converting GPS time to UTC is described in [10]. The public domain version of all data will use UTC time stamps.

Summary

An application of AV sensors and accurate, small-footprint GPS rovers has been used at Mcity to collect a novel combined dataset of high resolution camera data, vehicle position data, sensor data (lidar, radar), as well rover position files via independent differentially corrected measurements, to comprise a unique dataset for verification. The types of data and storage volumes are quantified in Table 2. These data can be leveraged to evaluation algorithms ranging from sensing, perception, detection, tracking, prediction, and classification.

An example trajectory from Act II, Scene II, in which the Ego vehicle traversed State street north and south, in Figure 10, shows the diversity of interactions between the

TABLE 2 Statistics for data collected in May 2017. For estimating data size, the collection in September covers approximately 2 hours

Data Type	Elapsed Time	Size	Timebase and Coordinate Frame
LiDAR	480 min (8 hours)	460 GB	Local UTC
Radar	480 min (8 hours)	300 MB	Local UTC
Camera	480 min (8 hours) × 7 cameras	2 TB	Local UTC, vehicle body frame
Vehicle Trajectory	480 min for ego car; up to two additional AV per scenario	25 MB	UTC, Global GPS
Rover Data	186 rover files	58 MB	UTC, Global GPS
UAV Video	29 files	74 GB	UTC, relative (not geo-referenced)

FIGURE 10 The locations of multiple actors, including the Ego vehicle, and multiple cyclists and pedestrians are plotted to show the variety of interactions represented in the data set.

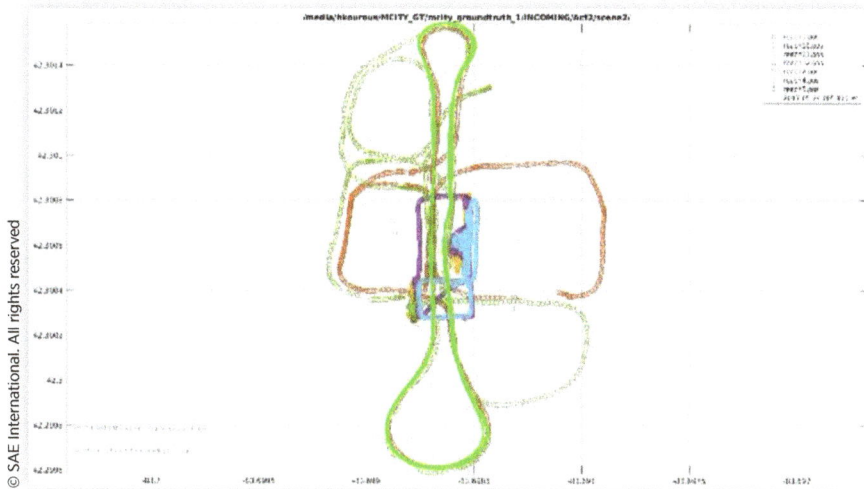

participants. This provides many opportunities to evaluate the accuracy of algorithms for detection, tracking, and classification.

The RTK units achieve fix or float quality correction better than 99% or the time, between 65% and 85%, at float quality (less than 1 meter of error) and 15% to 35% at integer (fix) quality (less than 10 cm of error.) In some cases, an example in Figure 11, shows a track with most of the corrected accuracy at less than 10 cm.

FIGURE 11 The position estimates for rover 4 from Act II, Scene 2, Take 2, are plotted on a tile from Google Earth. Green icons are used for estimates meeting integer (fixed) ambiguity resolution, and yellow for float. The Reference Position, where the base station was located, is shown just right of center in the image above.

Contact Information

Helen Kourous
Ford Motor Company
2000 Rotunda Drive, Dearborn MI
hkourous@ford.com

Jakob Hoellerbauer
Ford Motor Company
jhoeller@ford.com

William Buller
Michigan Technological University MTRI
3600 Green Court
Ann Arbor, MI 48105
wtbuller@mtu.edu
Sensor data and descriptive documentation will be available at following website http://mtri.org/multisensorevaluation. html

Acknowledgments

This work was funded by a research grant from Ford Motor Company.

The authors wish to thank Mcity for working with us on this complex endeavor. We thank the large support team from Ford: Sharath Nair, Wayne Williams, Kevin Walker, Codrin Cionca, Peng-yu Chen, Sangjin Lee, Matthew Warner, Thaddeus Townsend, James, McBride, and Zhiyuan Zuo. From Michigan Tech Research Institute: Melanie Feen, Bridget Taylor, Patrick McFall, Andrew Jurasek, Kyle Schwiebert, Jeremy Graham, Alice Eliot; Brian Wilson, Benjamin Hart, Michael Billmire, and Richard Dobson (drone pilot). This large cast served as drivers, data collection engineers, pedestrians, drone spotters, car parkers, and a myriad of other necessary tasks to make this happen.

Definitions/Abbreviations

ACC - Adaptive Cruise Control
AV - Autonomous Vehicle
LiDAR - Light Detection and Ranging
LCM - Lightweight Communications and Marshalling
PPP - Precise Point Positioning
RADAR - Radio Detection and Ranging
RTK - Real-Time Kinematic.
UTC - Coordinated Universal Time (English), Temps Universel Coordonné (Francais)

References

1. Benenson, R., Omran, M., Hosang, J., and Schiele, B., "Ten Years of Pedestrian Detection, What Have We Learned?" Lecture Notes in Computer Science (Including Subseries Lecture Notes in Artificial Intelligence and Lecture Notes in Bioinformatics), 2015.

2. Chakraborty, A., Stamatescu, V., Wong, S.C., Wigley, G. et al., "A Data Set for Evaluating the Performance of Multi-Class Multi-Object Video Tracking," arXiv preprint arXiv:1704.06378, 2017.

3. Dollár, P., Wojek, C., Schiele, B., and Perona, P., "Pedestrian Detection: A Benchmark," *IEEE Conference on Computer Vision and Pattern Recognition, CVPR 2009*, IEEE, June 2009, 304-311.

4. Piotraschke, H.F. and Optimal System DE, "RTK für Arme-Hochpräzise GNSS-Anwendungen mit den kostengünstigsten Trägerphasen-Rohdatenempfängern," GIL Jahrestagung, 2013, 271-274.

5. Vasileios Psychas, D. and Delikaraoglou, D., "Accuracy Improvement Techniques in Precise Point Positioning Method Using Multiple GNSS Constellations," *EGU General Assembly Conference Abstracts*, April 2016, Vol. 18, 3231.

6. Takasu, T., "PPP Ambiguity Resolution Implementation in RTKLIB v 2.4.2 Implementation in RTKLIB v 2.4.2," *PPP-RTK & Open Standards Symposium*, Frankfurt, Germany, 2012.

7. Dunning, D., "What Is Difference between RTK Fix & RTK Float?" Sciencing, April 25, 2017, https://sciencing.com/difference-between-rtk-fix-rtk-float-12245568.html.

8. Li, T. and Wang, J., "Some Remarks on GNSS Integer Ambiguity Validation Methods," *Survey Review* 44, no. 326 (2012): 230-238.

9. DJI Mavic Pro, https://www.dji.com/mavic/info.

10. Klepczynski, W. J., GPS for Precise Time and Time Interval Measurement, *Global Positioning Systems: Theory and Applications* (1996), 2.

CHAPTER 4

5

Integrating STPA into ISO 26262 Process for Requirement Development

Dajiang Suo, Sarra Yako, Mathew Boesch, and Kyle Post
Ford Motor Company

Developing requirements for automotive electric/electronic systems is challenging, as those systems become increasingly software-intensive. Designs must account for unintended interactions among software features, combined with unforeseen environmental factors. In addition, engineers have to iteratively make architectural tradeoffs and assign responsibilities to each component in the system to accommodate new safety requirements as they are revealed. ISO 26262 is an industry standard for the functional safety of automotive electric/electronic systems. It specifies various processes and procedures for ensuring functional safety, but does not limit the methods that can be used for hazard and safety analysis. System Theoretic Process Analysis (STPA) is a new technique for hazard analysis, in the sense that hazards are caused by unsafe interactions between components (including humans) as well as component failures and faults. Otherwise stated, STPA covers the safety analysis of system malfunctions as well as the safety of the intended function (SOTIF), in addition to Functional Safety..

This paper introduces a process map with a complete meta-model based on Systems Model Language (SysML) to support the integration of STPA into the functional safety process based on ISO 26262. In particular, the paper illustrates how STPA can help evaluate safety and other system-level goals with ASIL classifications from ISO26262's recommended Hazard Analysis and Risk Assessment (HARA). The meta-model can be also used to provide guidance on making architectural decisions in order to create functional safety requirements. To make the process map applicable to different functional safety processes adopted by OEMs, tool support is required. Guidelines on how to develop visualization tools based on the meta-model are given.

CITATION: Suo, D., Yako, S., Boesch, M., and Post, K., "Integrating STPA into ISO 26262 Process for Requirement Development," SAE Technical Paper 2017-01-0058, 2017, doi:10.4271/2017-01-0058.

Introduction

Developing requirements for automotive electric/electronic systems is challenging for several reasons. First, engineers have to deal with not only safety-related goals early in the concept phase, but also other system-level goals such as performance and security that decide stakeholders' satisfaction with new products. Often, architectural tradeoffs have to be made before creating detailed requirements [1]. Second, traditional hazard analysis techniques that deal with hardware failures are hard to use in complex software-intensive systems [3] [15]. Unintended interactions among software features contribute to off-nominal scenarios even if the system operates as designed. Third, the modeling and tool support for system engineering activities often vary from department to department, making system integration difficult in the sense that models and requirement artifacts differ among various processes.

To deal with these challenges, the International Organization for Standardization (ISO) extended the general functional safety standard IEC 61508 to a domain-specific standard of functional safety for the automotive industry-ISO 26262 [2]. It specifies various processes and procedures for ensuring functional safety, but does not limit the methods that can be used for hazard and safety analysis. In particular, its concept phase prescribes the process for "identifying a comprehensive list of hazards and causal factors in order to support the development of safety requirements" [4]. STPA is a new technique for hazard analysis in the sense that hazards are caused by unsafe interactions between components (including humans) as well as component failures and faults. Previous work on the use of STPA in the automotive domain includes the application of STPA for identifying hazards in an ISO 26262 compliant framework and the corresponding modeling and tool support.

Hommes [4] suggests that STPA be used in the concept phase in ISO 26262 with other hazard analysis techniques for identifying a comprehensive list of hazards and developing functional safety requirements. Mallya et al. [5] shows how "STPA can be used in an ISO 26262 compliant process" in which hazard analyses based on STPA are extended to include risk assessment. Thomas et al. [6] [7] propose an integrated approach in which hazard analysis and requirement generation can be conducted in parallel and iterative process. Although not specific to ISO 26262, this approach provides guidance and feedback for engineers to make critical design decisions during early concept development.

For the modeling and tool support for hazard analysis and requirement generation based on STPA, software prototypes and tools have been developed. Suo [13] illustrates a proof of concept tool for supporting formalized STPA Step-1 and automatic requirement generation. Abdulkhaleq develops an open-source platform [8] that allows engineers to perform STPA for requirement development and test case generation. The Safety Hazard Analysis Tool (SafetyHAT) [9] is developed by Volpe, the National Transportation Systems Center, to support STPA and customization for applications in different transportation systems. Although these efforts facilitate the use of STPA in the automotive industry, to the best of the authors' knowledge, modeling and tool support are still lacking in the following three aspects:

- Augmenting STPA to consider safety as well as other system- level properties such as customers' experience or cyber-security threats early in the concept development.

- Verifying functional safety requirements by detecting conflicts between safety constraints automatically.

- Providing guidance for assigning responsibilities to system components for defining preliminary architecture.

This paper introduces a process map for integrating STPA into the functional safety process defined by ISO 26262. In particular, it illustrates how to use STPA as the Hazard Analysis (HA) activity in an ISO 26262 compliant process for establishing safety and other system-level goals and to couple these to ASIL classifications that are based on current Risk Assessment (RA) methods to complete the HARA objectives of ISO 26262 Part 3. The process map described herein also provides guidance on making architectural decisions in order to accommodate functional safety requirements. To make the process map applicable to different functional safety processes adopted by OEMs, STPA extensions to the Systems Model Language (SysML) have been defined to facilitate using STPA with modeling and support tools. Guidelines on how to use the proposed process map are illustrated through a case study on an automated driving system. Figure 1 gives an overview of the integration work related to hazard analysis techniques in ISO 26262. All the work is based on ISO 26262-"Functional Safety Standard for Automotive Electric/ Electronic systems" [2].

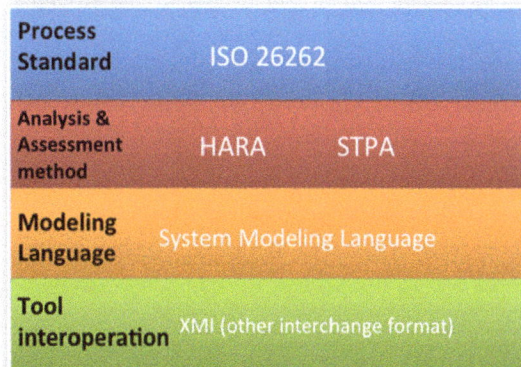

FIGURE 1 An overview of the integration work.

Process Map for STPA Integration

ISO 26262

ISO 26262 is the standard for functional safety of electrical and electronic systems in vehicles and an adaptation of functional safety standard IEC 61508 [20] specific to the automotive industry [2]. It "prescribes a system engineering process for safety engineering" [18].

Three clauses in the concept phase of ISO 26262 that are related to the development of functional safety requirements are considered for integrating STPA [16]. Although revisions have been made since the publication of the preliminary standard in 2011, reviews and assessments of ISO 26262 in [16] are still valid. Processes for developing technical safety requirements are omitted as they involve design details of components.

- Item definition: A system or subsystem that achieves a function at the vehicle-level.

- Hazard Analysis and Risk Assessment (HARA): HARA is used to identify hazards and appropriate reduction of risk captured by safety goals. Engineers first brainstorm possible hazardous events by deciding whether malfunctioning behaviors of a given item can cause hazards under various environmental conditions. The severity (S), exposure (E) and controllability (C) of the related hazardous event are assessed. Automotive Safety Integrity Level (ASIL) classification (A-D) is then assigned to each safety goal based on a fixed table mapping of {S, E, and C} to ASIL. For each hazardous event rated A-D, a safety goal (SG) is developed.

- Functional Safety Concept: The purpose of this step is to create functional safety requirements, at the concept level, to achieve safety goals developed in the HARA.

Although it gives a structuralized procedure for identifying hazardous events, the ISO 26262 standard provides limited guidance on specific hazard classification and mitigation.

STPA

STPA [3] is a new hazard analysis technique for complex systems in the sense that hazards are caused by unsafe interactions between components (including humans) as well as component failures and faults. STPA is based on STAMP-an accident causality models based on system and control theory [10]. Safety constraints, hierarchical control structure and process models are three basic concepts in STAMP. Accidents occur because safety constraints are not enforced by the control structure. Each "controller," whether human or computer, has a process or mental model for deciding whether or not to issue a control command to the process being controlled.

There are two steps in STPA, as shown in Figure 2. As an extension to Step 1, undesired control actions (UCA)* are identified and classified into four types:

- A control action required is not provided or not followed.

- An undesired control action is provided.

- A control action is provided too early or too late.

- A control action is stopped too soon or applied too late.

After UCAs are identified, Step 2 identifies causal factors and scenarios that potentially lead to UCAs. These are based upon the accident causality model described in STAMP.

It is worth mentioning that STPA can be extended beyond safety to other high-level system goals such as security [17] or performance.

Process Map for Creating Functional Safety Requirement

This paper proposes a process map that couples STPA with the functional safety process in ISO 26262, as shown in Figure 3. The map shows how meta-models (center) described by System Modeling Language (SysML) can facilitate the use of STPA (right) in the

FIGURE 2   Procedures for system theoretic process analysis.

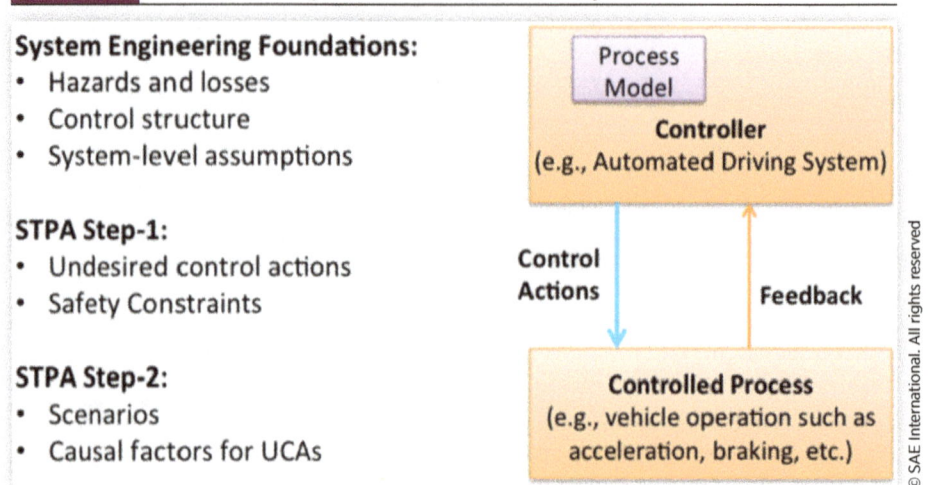

---

* UCA described in this paper is different from the standard definition [3] as other high- level goals of the system are considered.

FIGURE 3 Process map for integrating STPA with ISO 26262 process.

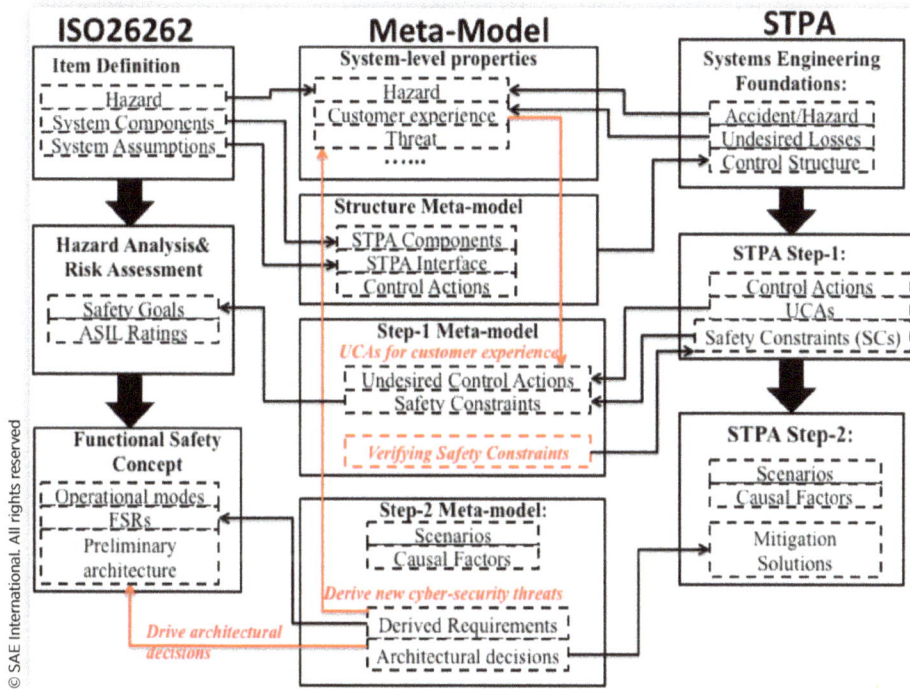

concept phase (left) in ISO 26262 for developing functional safety requirements and making architectural decisions.

The meta-model (top center) first takes as inputs hazard information as well as other system-level properties such as customer experience that are used to identify undesired control actions in STPA Step-1, as indicated by the red arrow (downward) in Figure 3. Also, new security threats can be derived as system-level properties if there exists cyber vulnerabilities in causal factors found in STPA Step-2, as indicated by the upward (red) arrow in Figure 3. The meta-model can also provide assistance for deriving the control structure as systems engineering foundations based on system components and assumptions from the item definition in ISO 26262. The second role of the meta-model is to STPA to complement HARA. With tool support for STPA Step-1 based on the meta-model, engineers can create and verify safety constraints to be associated with ASIL rated safety goals from risk assessments. The third use of the meta-model is to provide guidance for creating functional safety requirements that drive architectural decisions, as shown at the bottom center in Figure 3.

Modeling and Tool Support

Intro to SysML

The OMG System Modeling Language (SysML) is a graphical modeling language that provides support for the development of complex systems including specification, analysis, design and verification [11]. It includes three types of diagrams, each of which represents a specific aspect of the system, including behavior diagram, requirement diagram and structure diagram. A block is the basic unit that structurally describes hardware, software, facilities, personnel and external entities in the environment in the structure diagram. For this project, SysML is used as the base language that was extended to capture the control structure for STPA. The SysML requirement diagram at the highest level relates Safety Goals to safety requirements and other non-safety requirements derived from analysis based on STPA.

Meta-Model for Hazard Analysis & Requirement Generation

As shown in Figure 1 and Figure 3, modeling support (from either open-source or commercial tools) is necessary to make the process map scalable to complex automotive systems and usable by different engineering teams. This paper describes a meta-model based on a modeling environment that supports Systems Modeling Language so that safety constraints and requirements derived in STPA can be mapped into the elements that support functional safety processes in ISO 26262, as shown in Figure 4, Figure 5 and Figure 6.

Figure 4 illustrates how a control structure (right) can be built by using the meta-model (left) that defines system components and links. In addition to components in STPA, new elements are added into the model-model for visualization. For example, the "Control Action" block travels on the "command" link, and thus an association link connects them.

Figure 5 (left) gives the meta-model that can couple hazard analyses based on STPA with the process in ISO 26262. The upper part shows how visualization tools (in table forms) can be built in order to associate UCAs and safety constraints from STPA Step-1 results with safety goals and ASIL ratings developed in HARA. Engineers can then leverage the traceability between UCAs and scenarios and causal factors to create corresponding functional safety constraints, as shown on the bottom.

In addition to modeling support for STPA integration with ISO 26262, this paper also proposes a meta-model for verifying safety requirements and constraints created

FIGURE 4 Meta-model for building control structure based on item definition in ISO 26262.

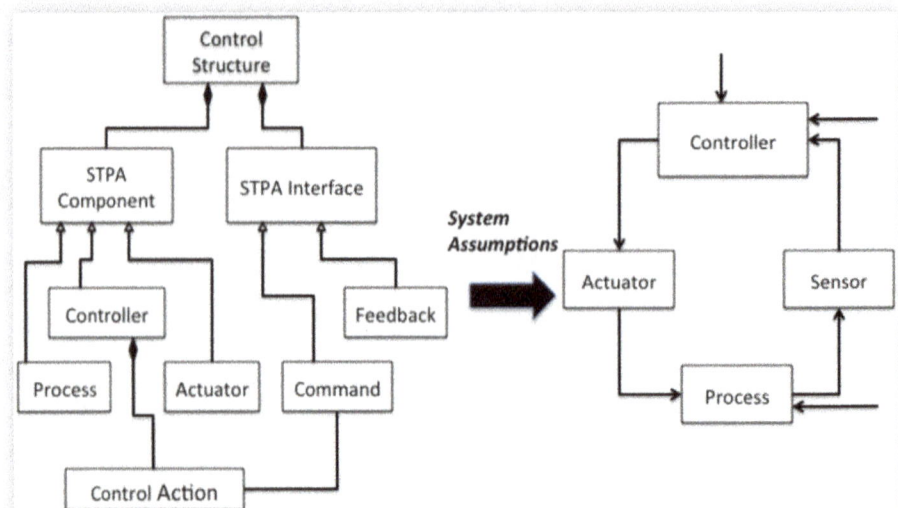

FIGURE 5 Meta-Model for supporting STPA Step-1 and Step-2.

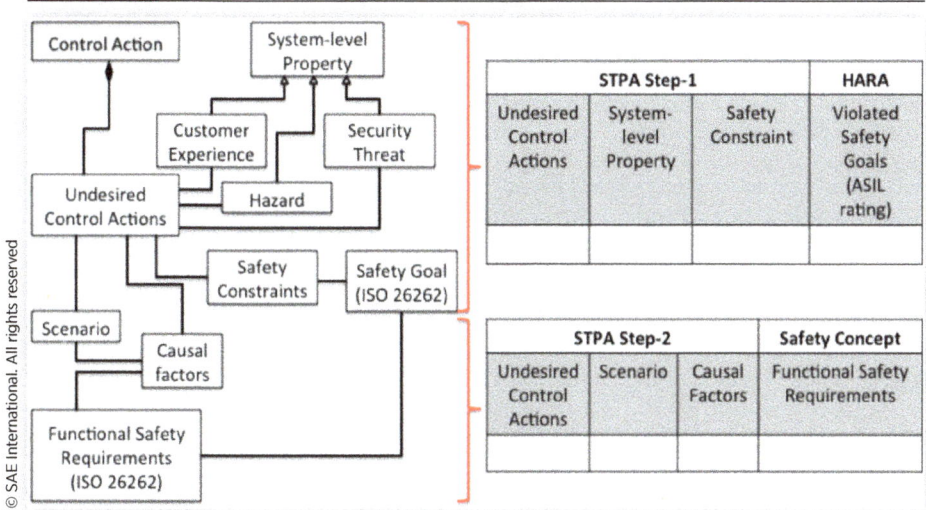

STPA Step-1			HARA
Undesired Control Actions	System-level Property	Safety Constraint	Violated Safety Goals (ASIL rating)

STPA Step-2			Safety Concept
Undesired Control Actions	Scenario	Causal Factors	Functional Safety Requirements

FIGURE 6 Meta-Model for verifying safety constraints.

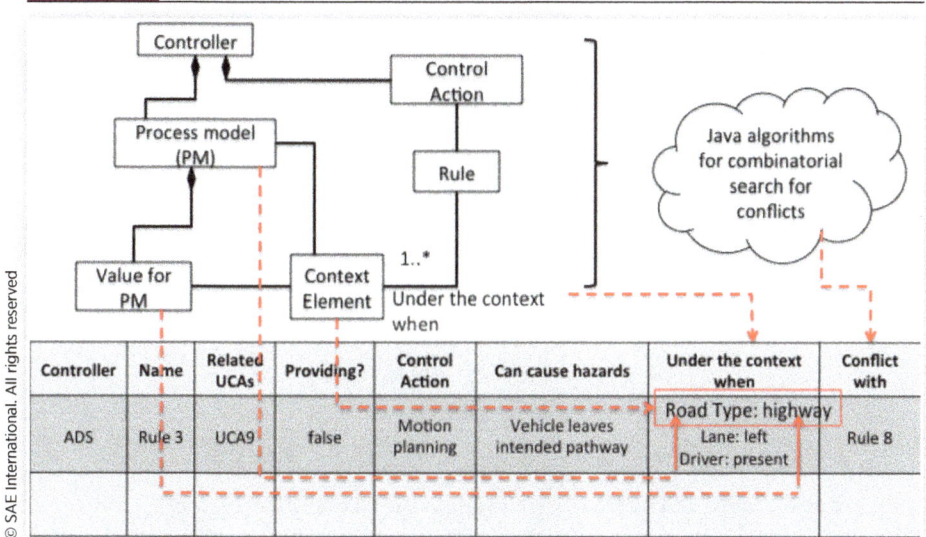

Controller	Name	Related UCAs	Providing?	Control Action	Can cause hazards	Under the context when	Conflict with
ADS	Rule 3	UCA9	false	Motion planning	Vehicle leaves intended pathway	Road Type: highway; Lane: left; Driver: present	Rule 8

in the functional safety process, as shown in Figure 6. This meta-model builds on two theoretical foundations:

- Process model. A control structure in STPA can have multiple controllers (e.g., automated or human controller) and each controller has a model of the process being controlled and the external environment [10], as shown in Figure 2. A process model may contain different variables that represent the system states or context. For example, the automated driving system can be in active mode, within the operational boundary, and on the correct route. Whether or not it is desired to provide the motion planning command is dependent on this context.

- A formal specification of hazardous control actions. Thomas [12] proposed a formal method to partially automate the process of identifying UCAs, identifying conflicts, and generating requirements. Specifically, a hazardous control action consists of four elements, including source controller, type (i.e., provided or not provided), control action and a context. With these four elements, engineers can define a "rule" which is the formalized UCA that contains a specific context. The rule is in a form that is close to natural language but machine readable for conducting combinatorial searches for determination of conflicting constraints.

Using the user interface based on the proposed meta-model in Figure 6, engineers can specify a "Rule" [12] [13] for UCAs. A "Rule" is a formalized way to specify contexts of a UCA [13]. The built-in Java algorithms in the plug-in developed by the author [13] can take these "Rules" as inputs and detect conflicts automatically. As shown in the last column of the table in Figure 6, "Rule 3" conflicts with "Rule 8", indicates a conflict between UCAs and corresponding safety constraints. A concrete example will be provided later in this paper.

System Engineering Foundations Based on Item Definition

To build the foundations for the system engineering process in ISO 26262, engineers take the inputs from item definition to construct a safety control structure with the modeling and tool support described in this paper. Those inputs include system-level hazards, system components and system assumptions.

As an example, three high-level hazards are chosen for this case study. G-1, a system-level goal related to customers' satisfaction or efficiency of the transportation system, is also included for illustrating the modeling support for dealing with different system-level properties.

- H-1: Getting too close to objects/terrain

- H-2: Vehicle leaves intended pathway

- H-3: Ingress/egress issue for rider

- G-1: Passengers' experience and traffic flow should not be disrupted by the operation of the automated driving system.

In addition, five assumptions are made on the automated vehicle:

- Passengers can request a ride service through the cloud infrastructure.

- The automated driving system and the road infrastructure support V2V or V2I communication through vehicular networks (e.g., Dedicated Short Range Communication - DSRC).

- The automated driving system can only operate within a prescribed operational boundary (e.g., time of the day, geographical boundary, weather conditions, etc.).

- After being activated, the automated driving system can handle all driving tasks appropriately.

- The human driver or passenger is able to interrupt the operation of the automated driving system.

The Modeling support for building system engineering foundations is shown in Figure 7. The SysML-based block diagram that represents system components (left) can be used to derive the control structure of the automated vehicle (right). This forms the basis of the example analyzed in the next section.

FIGURE 7 Safety control structure of the automated vehicle derived from item definition.

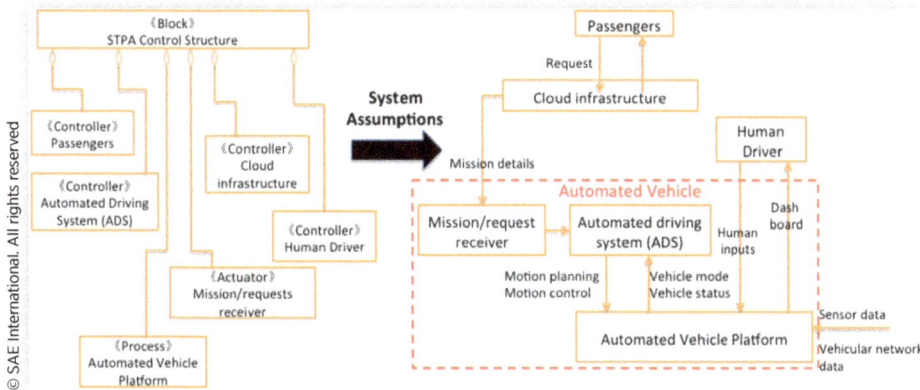

Integration of STPA Step 1 for Evaluating Existing Safety Goals

In STPA Step1, engineers identify undesired control actions to create initial (safety) constraints for the system. These constraints are then associated with existing Safety Goals from the HARA. If any Safety Constraint is not associated with an existing Safety Goal, then the HARA can be revised resulting in a new Safety Goal.

Several abstract safety-goals developed from HARA are as follows.

- SG-1: Prevent vehicle from leaving intended pathway.

- SG-2: Prevent vehicle from violating fixed separation distance from other vehicles.

- SG-3: Prevent human ingress/egress issues.

Two control actions-acceleration and motion planning-are chosen for the case study. Consider UCA 9 below. If not providing "motion planning" can cause any of the hazards defined above, a safety constraint should be created.

- UCA 9: Not providing motion planning is undesired if the automated driving system is in active mode, the vehicle is in route and within operational boundary [H-1, H-2].

- Safety Constraints 9: The automated driving system should provide motion planning if is in active mode, the vehicle is in route and within operational boundary.

Table 1 shows how safety goals with ASIL ratings from HARA can be assigned to each UCA and its corresponding safety constraint. The first column includes a set of UCAs for acceleration and motion planning commands, while the third column shows safety constraints derived for each UCA. For a given row, if any Safety Constraint is not associated with an existing Safety Goal, then the HARA can be revised, resulting in a new Safety Goal.

As an example of how STPA can be extended to deal with non-safety goals, consider a situation where the automated driving system does not provide the acceleration command (the 6th row in Table 1) when the automated vehicle is merging into a fast (left) lane on the highway, and its current speed is too low to smoothly merge into the traffic flow.

TABLE 1 Assigning ASIL rated safety goals to safety constraints

	STPA Step-1			HARA	
Undesired Control Actions	**System-level Property**	**(Safety) Constraints**		**Violated Safety Goals (derived from HARA)**	**ASIL (derived from HARA)**
UCA1: ADS Providing <u>Acceleration</u> command is undesired if the distance to the lead vehicle is less than the predefined threshold	H-1	SC1: ADS must not provide <u>Acceleration</u> command when the distance to the lead vehicle is less than the predefined threshold.		Prevent vehicle from violating fixed separation distance from vehicles	B
......					
UCA9: ADS Not providing <u>motion planning</u> command is undesired if the ADS is in active mode, the vehicle is en route and within operational boundary	H-1, H-2	SC9: ADS must provide <u>motion planning</u> command when in active mode, the vehicle is en route and within operational boundary.		Prevent vehicle from leaving intended pathway Prevent vehicle from violating fixed separation distance from vehicles	B B
Does Providing or Not Providing <u>Acceleration</u> or <u>motion planning</u> command violates G-1?	G-1				

New UCA derived from G-1

UCAX: ADS Not providing <u>Acceleration</u> command when vehicle is merging into the fast (left) lane on the highway	G-1	Non-safety related		Non-safety related	N/A

This is undesired (although not necessarily unsafe) because vehicles may need to brake and cannot maintain their desired speed that their passengers expect, thus violating G-1. Table 1 (last row) gives the newly derived UCA-ADS not providing <u>Acceleration</u> command is undesired when the vehicle is merging into the fast lane on the highway.

After identifying UCAs and creating corresponding safety constraints, it is necessary to verify Step 1 results to ensure that there are no conflicts between safety constraints or requirements. To help engineers finish this task, the authors develop a prototype user interface based on the meta-model in Figure 6. Table 2 gives an example of a conflict

TABLE 2 Detecting requirement conflicts automatically for verifying safety constraints

#	Name	Related UCAs	Providing?	Control Action	Can cause hazards	Under the context when	Conflict with
1	Rule 3	UCA9	false	Motion planning	Vehicle leaves intended pathway	Road Type: highway Lane: left Human Driver: present	Rule 8
2	Rule 1	UCA10	true	Motion planning	Ingress/egress issue for rider	Road Type: City street Lane: left Human Driver: present Weather: snow	
3	Rule 2	UCA10	true	Motion planning	Vehicle is operating without human driver	Time: midnight Human Driver: not present	
4	....	....		Motion planning			
5	Rule 8	UCA10	true	Motion planning	Getting too close to objects	Road Type: highway Weather: storm	Rule 3
	....	....		....			

existing between two safety constraints derived from UCAs. As can be seen, "Rule" 3 conflicts with "Rule 8", that is, Rule 3 requires ADS to provide motion planning when the automated vehicle is moving in the left lane on the highway and the driver is present, while Rule 8 asks ADS not to provide motion planning when the vehicle is moving on the highway under extreme weather conditions (e.g., storm).

Integration of STPA Step 2 for Creating Functional Safety Requirements

After creating safety constraints for validating safety goals for the <u>motion planning</u> and <u>acceleration</u> commands, engineers can perform STPA Step-2 to:

- Create functional safety requirements for each safety goal.

- Assign responsibilities to system components for defining the preliminary architecture of the automated vehicle.

The meta-model provides the traceability necessary for performing these tasks. Consider UCA 9 -Not providing <u>motion planning</u> is undesired if the automated driving system is in active mode, while the vehicle is en route and within operational boundary. Causal scenarios and factors related to this UCA include:

Scenario 9: The automated vehicle is moving on the highway and a lane-change is needed because of newly detected events or obstacles. But it does not change lanes because the automated driving system does not provide motion planning in time. The automated controller believes that the vehicle is not in the correct route or within the operational boundary.

- Causal factor 9.1: mission details are incomplete.

- Causal factor 9.2: unintended mode changes from autonomous drive mode to manual mode.

- Causal factor 9.3: automated driving system believes that the automated vehicle is not en route because map data was modified by unauthorized access.

- Causal factor 9.4: automated driving system believes that the automated vehicle is not in the operational boundary because map data (e.g., route direction) or timing information (e.g., clock) was modified by unauthorized access.

While the first two causal factors are recognized as safety concerns, the third and fourth are related to cyber vulnerabilities in the given architecture described before. This can be understood by an example involving the "reversible lane" that is designed to allow the traffic to move in either direction based on certain conditions [19]. It is designed to avoid traffic jams during rush hours, as shown in Figure 8. As can be seen, the yellow lane in the middle is designed to be "reversible," and its traffic direction depends on the time of a day. Traffic moves from east to west (left) in the morning while moving in the opposite direction (right) in the evening. Currently, the lane direction is indicated by control signals LED signs or physical separation. In the future, when automated vehicles support V2I communication, such design can be achieved if the automated driving system receives route information from road units through vehicular network, such as DSRC.

One problem is that V2I communication is subject to cyber-attacks. For the vehicle architecture that only allows autonomous vehicles to move within the "operational design domain" (e.g., time, geographical condition, etc.), an adversary who can modify the map data related to the highway direction or spoof the timing information (e.g., vehicle clock)

FIGURE 8 An example of UCA-9 caused by cyber vulnerabilities.

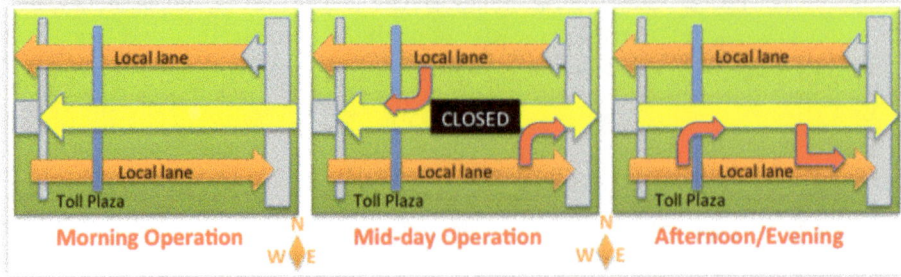

can easily compromise the normal operation of the vehicle. Consider that the vehicle is moving in the reversible lane from east to west, but the timing information that the automated driving system receives suggests that it is in the evening. This is the case where the automated driving system finds itself "out of the operational design boundary" because it can only move the vehicle from east to west in the reversible lane during the morning, according to the traffic rules. To make things worse, if the automated driving system is not designed for moving the vehicle to a "safe" place when the vehicle is not en route or is out of operational boundary, it will not stop providing motion planning and control commands, putting the automated vehicle in a dangerous situation (no one is controlling the vehicle).

Based on the scenarios and causal factors identified above, functional safety requirements can be created to define the vehicle's architecture. Figure 9 illustrates how FSR 9.1 is created by considering the worst-case scenario. According to the system assumption that an automated vehicle can only operate within the operational boundary, ADS will stop providing motion-planning command under the scenario because of causal factors 9.3. Engineers may then decide to assign the responsibility of keeping the vehicle moving until it is in a "safe" zone to ADS, as illustrated by Architectural decision 1.

FIGURE 9 Traceability for creating FSRs and making architectural decisions.

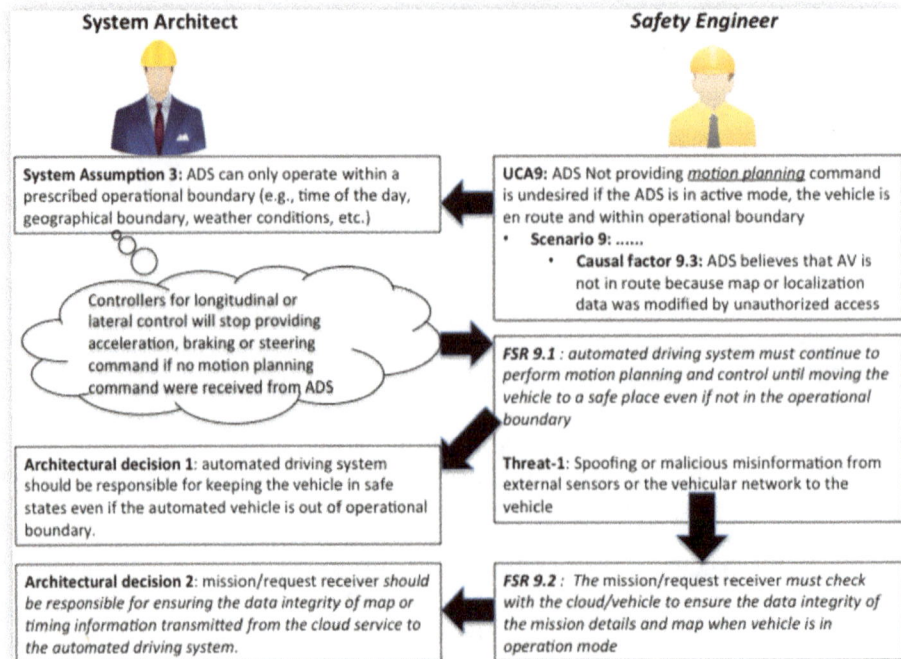

In addition to deriving functional safety requirements, scenarios and causal factors identified in STPA Step-2 also provide guidance in identifying cyber-security threats. For example, a cyber-security threat (Figure 9) related to scenario 9 and causal factors 9.3 and 9.4 can be

- Threat-1: Spoofing or malicious misinformation from external sensors or the vehicular network to the vehicle

   Although Threat-1 is framed as a security concern, it also has safety implications, as spoofing attacks on the map data or vehicular network can also result in vehicles violating minimum separation distance (H-1) or not following the intended pathway (H-2) that is originally planned by the automated driving system. Engineers can create FSR 9.2 to specify the required behavior of the mission/request receiver-the mission/request receiver must check with the cloud/vehicle to ensure the data integrity of the mission details and the map when the vehicle is in operation mode. Figure 9 only illustrates how STPA Step-2 results can be used to derive functional safety requirements and responsibility assignment, rather than specifying a user interface for performing this task.

   As guidelines for engineers when making architectural decisions, consider several guidance questions after creating functional safety requirements for the automated driving system.

- Given the system assumption that the automated driving system should only operate within its operational boundary (e.g., time, geographical and route conditions, weather conditions, etc.), which module should keep the vehicle safe when the automated vehicle is still moving but the automated driving system is deactivated because it is out of the operational boundary?

- If the automated driving system is responsible to move the vehicle to a "safe" place even if it is out of the operational boundary, what test cases are necessary to ensure the system is devoid of hazardous behaviors?

- If another controller is required to operate the vehicle when the automated driving system is not within the operational boundary due to unexpected events, what safety requirements should be created such that control commands from the new controller do not conflict with the old one's?

Consideration for Integration with Cyber Security Analysis

Safety assurance in ISO 26262 does not consider cyber security issues that are the focus of the cyber-security guidebook for automotive systems-SAE J3061 [21]. One of the goals of providing the cybersecurity guidebook is to "evaluate threat analysis and risk assessment (TARA) methods using a simple approach to allow effective implementation across the automotive industry." [22] STPA can complement the threat analysis in the sense that it can find missing threats or evaluate existing ones, rather than substituting TARA.

   Figure 10 illustrates how Threat-1 derived from STPA Step-2 results. The case study on an automated driving system starts with high-level hazards and losses that must be prevented (i.e., two vehicles violating minimum separation distance) and two functions for the automated driving system-motion planning and acceleration, rather than explicitly considering potential threats or malicious attacks to the automated vehicle. After finishing Step 1 and Step 2 with all causal scenarios and factors identified for UCAs, engineers are able to create not only functional safety requirements, but also security requirements for system components (e.g., the automated driving system and the mission/request receiver). Also, for the message that is used by the automated driving system,

FIGURE 10 Relation diagram for functional safety, STPA and cyber-security.

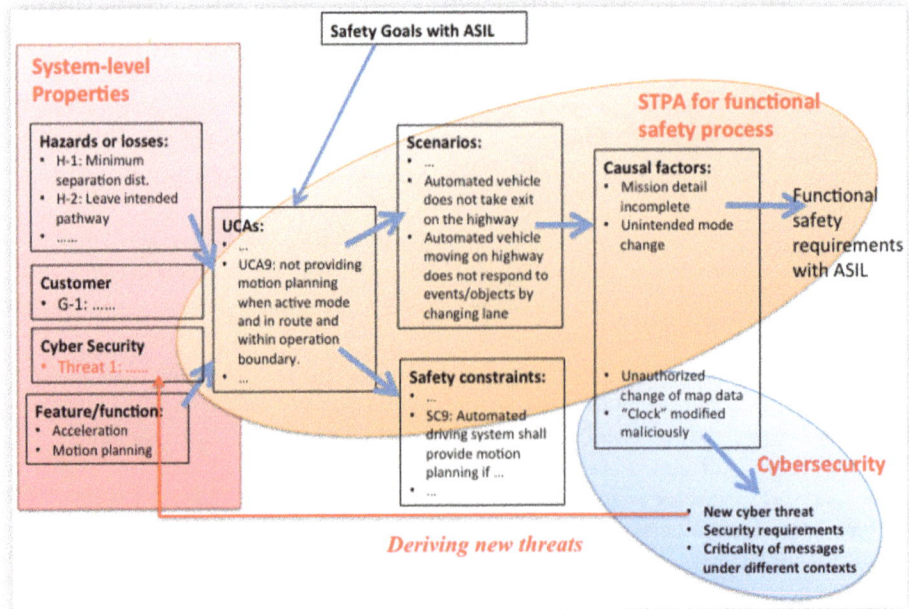

such as route direction in the map data or timing information, engineers can decide its criticality based on which function (e.g., motion planning) is using it and the ASIL rating of the safety goal for that function.

Summary/Conclusions

This paper describes a process map for integrating STPA into the functional safety process based on ISO 26262. Specifically, three steps in the process map are illustrated through a case study on an automotive system.

- System assumptions and components from item definition are used to form the system engineering foundations for STPA.

- UCAs identified and safety constraints created in STPA Step 1 are used to evaluate existing safety goals with ASIL ratings developed from HARA.

- Causal scenarios and factors for UCAs identified in STPA Step 2 help engineers create functional safety requirements and make architectural decisions.

For the modeling and tool support, a meta-model based on SysML is developed. In addition to enabling the integration of STPA with ISO 26262 process, the paper also shows how the meta-model can be used to deal with system-level properties other than safety early in the concept phase, i.e., identifying undesired control action related to customer experience in STPA Step-1 and derive new cyber security threats based on scenarios and causal factors from STPA Step-2. It is worth mentioning that STPA is an iterative process and can also be used to create technical safety requirements and generate test cases. But those aspects are not covered because the focus of the paper is on functional safety requirements.

Definition/Abbreviations

STPA - System Theoretic Process Analysis
SC - Safety Constraint
CF - Causal Factor
ADS - Automated Driving System

References

1. Flemming, C., *Safety-Driven Early Concept Analysis and Development* (Cambridge, MA, 2015).

2. International Standardization Organization, ISO 26262- 1:2011(en) Road vehicles - Functional safety - Part 1: Vocabulary. International Standardization Organization.

3. Leveson, N., *Engineering a Safer World* (Cambridge, MA: MIT Press, 2012).

4. Hommes, Q., "Safety Analysis Approaches for Automotive Electronic Control Systems," https://www.nhtsa.gov/sites/nhtsa.dot.gov/files/2015sae-hommes-safetyanalysisapproaches.pdf.

5. Mallya, A., Using STPA in an ISO 26262 Compliant Process," *Computer Safety, Reliability, and Security: 35th International Conference, SAFECOMP 2016*, Trondheim, Norway, Springer, September 21-23, 2016.

6. Thomas, J., Sgueglia, J., Suo, D., Leveson, N. et al., "An Integrated Approach to Requirements Development and Hazard Analysis," SAE Technical Paper 2015-01-0274, 2015, doi:10.4271/2015-01-0274.

7. Placke, S., Thomas, J., and Suo, D., "Integration of Multiple Active Safety Systems Using STPA," SAE Technical Paper 2015-01-0277, 2015, doi:10.4271/2015-01-0277.

8. Abdulkhaleq, A., and Wagner, S., "XSTAMPP: An eXtensible STAMP Platform as Tool Support for Safety Engineering," 2015.

9. Becker, C. and Hommes, Q., Transportation Systems Safety Hazard Analysis Tool (SafetyHAT) User Guide (Version 1.0). No. DOT-VNTSC-14-01, 2014.

10. Leveson, N., "A New Accident Model for Engineering Safer Systems," *Safety Science*, 42, no. 4 (2004): 237-270.

11. Object Management Group, The OMG System Modeling Language Version 1.4 specification, 2015, http://www.omg.org/spec/SysML/1.4/

12. Thomas, J., "Extending and Automating a Systems-Theoretic Hazard Analysis for Requirements Generation and Analysis," Ph.D dissertation, Cambridge, MA, 2013.

13. Suo, D., "Tool-Assisted Hazard Analysis and Requirement Generation based on STPA," Master thesis, Cambridge, MA, 2016, http://hdl.handle.net/1721.1/105628

14. Suo, D. and Thomas, J., "An STPA Tool," *3rd STAMP/STPA Conference*, Cambridge, MA, 2014.

15. Leveson, N., "Completeness in Formal Specification Language Design for Process-Control Systems," *Proceedings of the Third Workshop on Formal Methods in Software Practice, ACM*, 2000, 75-87.

CHAPTER 5

16. Van Eikema Hommes, Q., "Review and Assessment of the ISO 26262 Draft Road Vehicle-Functional Safety," SAE Technical Paper 2012-01-0025, 2012, doi:10.4271/2012-01-0025.

17. Young, W. and Leveson, N.G. "An Integrated Approach to Safety and Security Based on Systems Theory," *Communications of the ACM* 57, no. 2 (2014): 31-35.

18. Hommes, Q., Assessment of safety standards for automotive electronic control systems, (Report No. DOT HS 812 285), National Highway Traffic Safety Administration, Washington, DC, June 2016.

19. https://en.wikipedia.org/wiki/Reversible_lane.

20. Bell, R., "IEC 61508: Functional Safety of Electrical/Electronic/ Programme Electronic Safety-Related Systems: Overview," *Computing & Control Engineering* 11, no. 1 (1999): 5/1-5/5.

21. SAE International Surface Vehicle Recommended Practice, "Cybersecurity Guidebook for Cyber-Physical Vehicle Systems," SAE Standard J3061™, Iss., January 2016.

22. SAE International, "SAE Committee Busy Developing Standards to Confront the Cybersecurity Threat," Automotive Engineering Magazine article, http://articles.sae.org/13809/.

23. Ujiie, R., "Using STPA in the Design of a New Manned Spacecraft," The 2nd STAMP Workshop, Cambridge, MA, 2013.

CHAPTER 6

Hazard Analysis and Risk Assessment beyond ISO 26262: Management of Complexity via Restructuring of Risk-Generating Process

Oleg Lurie and Joseph Miller

ZF - TRW

The automotive world is getting ready to embrace the automated driving (AD), while advanced driver assistance systems (ADAS) increase their authority in the control over the vehicle. It is necessary to guarantee system safety of the AD/ADAS application, which includes both "classic" functional safety according to ISO 26262 and specific areas like Safety of the Intended Functionality (SOTIF) and others. However, safety remains safety, that is, absence of unreasonable risk. All safety activities within a project, therefore, need to have their source in a Hazard Analysis and Risk Assessment (HARA), encompassing all relevant aspects, including operational situations, description of functionality and other parameters.

Already from the description a HARA for an AD/ADAS is going to be a complex task. Here we demonstrate an approach for complexity management of HARA for an ADAS system. A manageable overview of potential hazards resulting from malfunctions as well as from external causes was obtained and SOTIF validation goals were defined.

CITATION: Lurie, O. and Miller, J., "Hazard Analysis and Risk Assessment beyond ISO 26262: Management of Complexity via Restructuring of Risk-Generating Process," SAE Technical Paper 2018-01-1067, 2018, doi:10.4271/2018-01-1067.

Introduction

According to ISO 26262-3, a HARA consists of situation analysis and hazard identification, classification of hazardous events, and the determination of the safety level. The safety level largely determines further stages of the safety lifecycle, up to the safety validation.

Safety of the Intended Functionality (SOTIF) is an extension of the safety lifecycle. While the functional safety covers the hazards caused by malfunctioning behavior (i.e. unintended behavior of the item with respect to its design intent), DPAS 21448 "Safety of the Intended Functionality" [1] describes a specific safety lifecycle addressing performance limitations of the intended behavior or by reasonably foreseeable misuse by the user. This lifecycle includes "SOTIF HARA". We consider the SOTIF HARA to be an extension to the HARA according to ISO 26262 [2].

The following features distinguish the HARA related to the SOTIF from the HARA described in ISO 26262-3:

- *Hazard analysis:* even though the severity and controllability estimations use the same scales, their determination are specific for SOTIF hazards.

- *Safety levels* are not specifically addressed. The term "acceptable risk", which is often used in the document, refers to the acceptability of severity and controllability (S0 and C0 evaluations respectively).

- SOTIF HARA includes the *specification of a validation target*. Evidently, specification of a validation target requires the method of validation also to be specified.

The latter point may be solved in different ways. DPAS 21448, Annex B, suggests real-world driving tests on public roads. Once the cars equipped with the system in validation have driven the defined number of kilometers, their statistics can be compared to the available statistics of human driving, considering given region and type of a hazardous event. The GAMAB principle may be applied here: if the automated function performs at least as safe as human drivers, it may be considered safe.

According to ISO 26262, hazard identification is based on the situation analysis, which in turn consists of operational situations and the operating modes for the system in question. That means, a HARA consists of multiple situations often different only in minor details. This approach generates situations which may be different from each other only in minor details. Fitting the driving statistics into such detailed state description is not always possible or may be too arbitrary. Besides that, the management of many situations is problematic due to its complexity.

The reference HARA used for this paper contains 20 operating states to be considered for an AD system, which are analyzed in 28 situations. Out of 560 possible combinations, 166 were selected for further considerations. Those were counter-imposed to 23 possible environmental conditions. 1080 resulting situations are selected for final analysis. After the situations are clustered and possible malfunctions are added, the final analysis spreadsheet contains 1867 lines, each containing a pair of situation and a hazardous event.

In the following chapter, a method for lowering the HARA complexity by re-design of the underlining risk-generation process is described. An example of application for this approach is demonstrated. Its viability, strong and weak points are discussed. Finally, conclusion and an outlook over the future work is presented.

SOTIF HARA and State Space Explosion

State space explosion problem is a term coming from computer science, specifically from the area covering formal model checking of algorithms. Essentially, state space explosion is a situation where the number of states a system may assume (and therefore the states which should be considered by the model-checker) grows exponentially with increasing numbers of the parameters to be considered. By way of analogy, we may speak of the state space of the HARA, meaning all potentially hazardous situations which need to be analyzed.

HARA Composition

The goal of the HARA is to analyze hazards present in the system with regard to the system's different use cases. Therefore, the potentially hazardous situations of the HARA are built out of the superposition of those two. Below, we analyze the process for generation of hazards and of the use cases, i.e. the "risk-generating" process.

HAZARDS

According to ISO 26262, the HARA is performed on an *item*, i.e. a function implemented on the vehicle level is analyzed. Hazards are expressed as functional failure modes of the item. The failure modes of a technical system may be evaluated by analysis of its composition and working principles.

In case of SOTIF, the function in question is normally a part of driver's responsibilities which is being taken over by an AD/ADAS. The definition of this function should include driver's task breakdown (as the human drivers do not consciously distinguish between e.g. control tasks in longitudinal and lateral dynamics) and the sub-tasks which used to be solved by the driver and now going to be automated. The failure modes to be considered here are the "failure modes" of the driver, i.e. possible failures in fulfilling a driving-related task, as driver is now out of the loop. [3]. HAZOP uses matrix consisting of guidewords and signal names to generate a set of possible hazards [4]. The generated set is then subjected to plausibility analysis. Implausible combinations of signals and guide-words are then excluded. However, according to our experience, it is rarely possible to exclude driver-related hazards identified by HAZOP upfront, as drivers are prone to very different mistakes in controlling their vehicles. Errare humanum est.

USE CASES

Use cases for HARA include operating states, driving situations and environmental conditions. Not all environmental conditions are relevant for all states, e.g. for a passive safety system (i.e. a system minimizing harm to vehicle occupants after crash) it does not matter, whether the road surface was slippery before crash or not. However, in case of AD/ADAS functions, the object of the analysis is the driver's behavior. It is hard to predict which parameters are influencing the driving and which are not, therefore most of the parameters shall be taken into account. Besides that, an AD/ADAS function should be analyzed in all driving situations, both in those where it is intended to be used and in those, where it is not.

HARA and the Hidden Semi-Markov Chain

Driving (as virtually any process) may be represented in the form of a transition diagram (see Figure 1). Furthermore, a Markov process can be built upon this diagram [5]. The process would describe the development of a hazardous event (e.g. an accident), i.e. precisely the events studied by the HARA.

The driver-dependent part of transitions shall be considered under "hazards" (more precisely, the "hazards" consider the situations when the driver or an ADAS/AD system is taking a wrong decision), while the driver-independent transitions shall be considered under "use cases". The probability of the use case constitutes the "exposure" (E) parameter of the ISO 26262-compliant HARA, while the probability of transition between hazards and "no accident" state defines "controllability" (C). Severity (S) is a property of the accident; it is not directly derivable from the semi-Markov chain pictured on the Figure 1.

The word "hidden" in the title of this section refers to the fact that there is no possibility to measure many of the probabilities and the rates of the chain directly. The major source of information on traffic statistics are the accident statistics, i.e. we are dealing here with a sample for which statistics are skewed by definition, as we would never know out of driving statistics how many people have experienced the hazard and managed to avoid the accident.

It is possible to calculate the distribution of the probability of the system being in each state of the semi-Markov chain, if the form of probability distributions for the transitions and its parameters are known. For the transition diagram presented on the Figure 1, the number of the transitions is the power of the following set:

$$Transitions = UC \times Hazards \cup Hazards \times Accidents \qquad (1)$$

Each transition is governed by a specific probability distribution, which may differ from other distributions not only in parameters, but also in its form.

The approach of this paper is to find another, smaller set of probability distributions by changing the transition diagram of the semi-Markov processes underlying the HARA, while still being able to extract the HARA-relevant information as well as to validate the results against traffic and accident statistics and to generate a suggestion with regard to the validation of the SOTIF.

FIGURE 1 Semi-Markov chain representing processes studied via HARA.

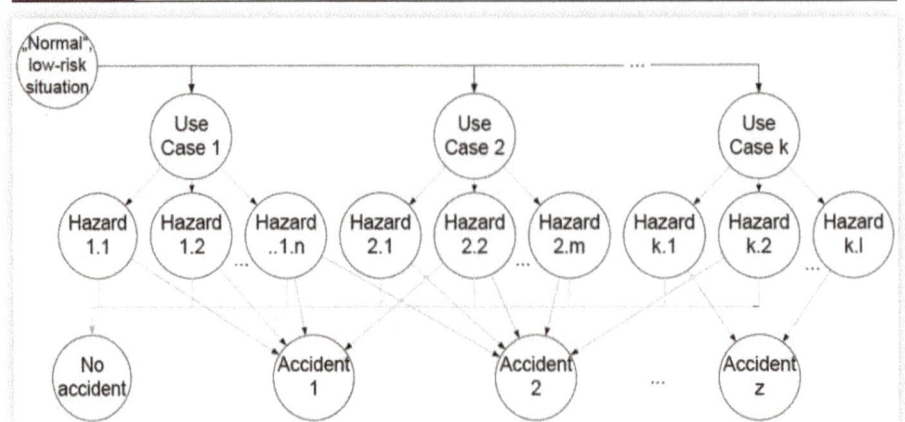

FIGURE 2 Model of road traffic risks.

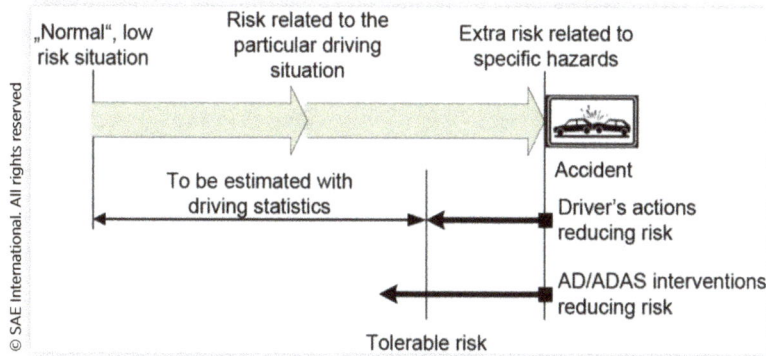

Restructuring of Risk-Generating Process

The road traffic is an extremely regulated area of human activity. It cannot be considered a Brownian motion. Road behavior of trained drivers features a combination of taught maneuvers under influence of the actual driving situation, which in turn is determined by the weather conditions, road type, behavior of other traffic participants and other parameters.

In order to minimize the number of the use cases, it makes sense to bound them to something easily identifiable and measurable. For AD/ADAS systems, it is beneficial to follow the approach proven by the process of teaching for human drivers, i.e. to benchmark the systems' performance in separate traffic situations. The choice of traffic situations as a milestone for risk development is substantiated by the relative accessibility of statistics on the use of different type of roads (cf. [6]). Furthermore, accident statistics often gives information on the driving situation preceding the accident, allowing easier consideration of driving situations w.r.t. the causes of the accidents. The usage of driving statistics obtained from human drivers is valid as long as the cars equipped with AD systems represent a minority of vehicles on the roads. As long as there is no surge in the amount of AD-equipped vehicles, the actual driving statistics reflects the benchmark for the driving safety.

A model implementing driving situations as important predictors of the risk is shown on Figure 2. According to the model, it is assumed that initially (e.g. when car is stopped and parked), there is a situation which can be considered low risk or background risk. From here, the risk increases depending on the driving situation. In stating so, we *qualitatively* compare risks related to e.g. driving in a multi-lane road with no traffic with those related to driving on a single-lane road with vehicle parked alongside and tense traffic, and conclude that the latter has "higher risk" associated with it than the former. Our goal, however, is to estimate the risk *quantitatively* via probability of an accident, and to determine the tolerable risk, which is then used as a target for AD/ADAS validation. Therefore, we need a semi-Markov process as discussed above, and we obtain it by way of modification of the chain depicted on the Figure 1, taking into account the risk model.

Figure 3 depicts the transition diagram for the semi-Markov chain which could be used for HARA. For the transition diagram presented on the Figure 3, the number of the transitions is the power of the following set:

$$Transitions = Situations \cup Situations \times Accidents \qquad (2)$$

FIGURE 3 Semi-Markov chain for HARA.

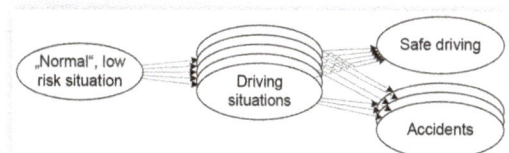

Comparing equations (1) and (2), it is safe to assume that the Figure 2 yields a semi-Markov process with considerably less transitions than the Figure 1.

Automatic Emergency Braking (AEB) Example

Markov Chain Solution

In order to illustrate the proposed method, a HARA for an AEB functionality will be studied.

Automatic Emergency Braking (AEB) detects an impending forward crash in time to avoid or mitigate the crash. These systems first alert the driver to take corrective action to avoid the crash. If the driver's response is not sufficient to avoid the crash, the AEB system may automatically apply the brakes to assist in preventing or reducing the severity of a crash [8]. AEB are mandated for some categories of trucks (see UN ECE R 131 or EU 661/2009). It is now being introduced also in passenger cars.

The HARA for AEB functionality is being performed under consideration of the two goals: first, ASIL shall be determined according to ISO 26262, and second, the driving distances for SOTIF validation shall be calculated. The HARA has been performed on the basis of the urban driving scenarios HARA (see [3]). The parameter of controllability by the driver was additionally validated by Monte-Carlo simulation [9].

SOTIF validation mileage was calculated via probabilistic approach.

The probabilistic approach included a stochastic process with multiple states modelled via state transition system as shown on Figure 3. For simplicity, all transitions rates were considered stationary; here we may speak of a Markov process instead of a semi-Markov process as considered above.

The sojourn probabilities in a Markov process are described by Chapman-Kolmogorov equation:

$$\dot{P} = \Lambda P \tag{3}$$

Where P is a vector representing sojourn probabilities and Λ is a transition matrix with an element λ_{ij} denoting transition rate between states i and j.

In the equation (3) we consider the probability against the number of driven kilometers $P(l)$, with $\dot{P} = dP / dl$.

Eq. (3) possesses an analytical solution [5]:

$$P_l = \exp(\Lambda l) P_0 \tag{4}$$

Where P_l is the sojourn probability vector after l kilometers of travel, and P_0 is the initial condition for the same. $P_0 = [1\ 0\ 0\ ...\ 0]^T$ by definition, as all vehicles always start from the "normal, low risk" state. Driving statistics defines l (average yearly mileage of one vehicle) and Pl (vector containing probabilities of different accidents happening in a year). Solving the eq. (4) for Λ will yield all transition probabilities between states depicted in Figure 3; distance to accident may be calculated as an inverse of the relevant transition rate:

$$MDTA_{ij} = 1 / \lambda_{Situation_I \rightarrow Accidents_J} \tag{5}$$

Eq. (5) shows that if Λ is known, it is easy to find the expected time to a specific accident in each driving situation. The validation can be performed focused on the driving situation via driving for $MDTA_{IJ}$ kilometers in a driving situation I and proving that no breach of safety has happened. A safety coefficient may be used as shown in PAS 21448 Annex D.

Eq. (4) has an analytical solution for Λ involving matrix logarithm. Matrix logarithm involves calculation of a sum of a row which converges only conditionally (in contrast, the row representing the matrix exponent always converges). Therefore, numerical solution was chosen involving random gradient descent to find Λ.

Regions of the Transition Matrix

The transition matrix Λ may be divided into the regions as shown by the eq. (6):

$$\Lambda = \begin{bmatrix} O_1 & D & O_2 \\ O_3 & O_4 & A \end{bmatrix} \tag{6}$$

Here D represents the probabilities of transition from the initial "low risk" state into the driving state, A represents the transitions between driving states and the accidents (including "safe driving" as non-accident), and O are zero matrices of the appropriate size. The D part is taken from the traffic data [7]. Our goal is to determine A fulfilling the Eq. (4).

The matrix A itself is also sparse, as transitions between driving situations and accidents are not arbitrary. Table 1 shows existing transitions (non-zero transition rates) in the matrix A.

Kinds of the accident are selected according to [7]:

1. Collision with another vehicle which starts, stops or is stationary
2. Collision with another vehicle moving ahead or waiting
3. Collision with another vehicle moving laterally in the same direction
4. Collision with another oncoming vehicle
5. Collision with another vehicle which turns into or crosses a road
6. Collision between vehicle and pedestrian
7. Collision with an obstacle in the carriageway
8. Leaving the carriageway to the right or left

Driving situation according to HARA (cf. [3]):

S1. Vehicle in motion at constant speed, in heavy traffic
S2. Vehicle accelerating, in heavy traffic
S3. Vehicle changing lanes on a multi-lane road with heavy traffic
S4. Vehicle turning, on an extreme curve or at a marked or unmarked intersection, with heavy driving

In the Table 1, both "X" and "P" denote a non-zero element of the matrix A. The difference here is assumed in the magnitude of the transition rates. "P" (meaning "potentially") marks transitions which are possible only under the condition of a breach of the traffic code. Consider "P" in row 1, column S1. This combination refers to a vehicle, being in motion with constant speed in heavy traffic, colliding with a vehicle driving off its parking spot. According to the traffic code, the driving off vehicle shall let the traffic pass before leaving the spot. For this example, we assume that transitions related to the breach of the traffic code happen at least one order of magnitude more rarely than the "normal" transitions (this assumption may be revised or completely ignored later, as it is not important for the method).

TABLE 1 Non-zero transitions in the part A of transition matrix Λ

Kind of Accident	S1	S2	S3	S4
1		P	P	
2		X	X	
3				X	P
4				P
5	P	P	P	P
6		X	X		X
7		X	X		X
8				P

TABLE 2 Median Distance to Accident (MDTA), thousands of km

Situation	S	MDTA, ×1000 km
Vehicle in motion at constant speed, in heavy traffic	S1	60
Vehicle accelerating, in heavy traffic	S2	75
Vehicle changing lanes on a multilane road with heavy traffic	S3	60
Vehicle turning, on an extreme curve or at a marked or unmarked intersection, with heavy driving	S4	6
Total	$\sum S_{1..4}$	201

Table 2 shows the median distances to accident depending on the driving situation. The data are obtained based on the transition rates of a Markov chain depicted on Figure 3, using Eq. (5).

The total mileage presented in Table 2 is comparable with the mileage calculated via alternative methods (cf. [9] where 153,000 km is the calculated MDTA for all driving states based on the US driving statistics).

Is There Another Way to Do It?

Alternatively, the validation distance may be calculated applying the formula from PAS 21448, Annex B (Eq. 7) directly.

$$D_{km} = \frac{NK}{A_v} C_a P \qquad (7)$$

Here N is the number of vehicle in the field; K - average annual mileage per vehicle (i.e. NK is the average mileage all vehicles travel per year); A_v is the amount of accidents of the relevant type per year (for AEB it would be rear- and front-end collisions); C_a is the assurance factor; and P is the probability of accident resulting from the given situations.

Calculation of P poses a specific challenge to this approach. It was suggested to use Monte Carlo modelling to determine P [10]. However, the modelling requires validation, which due to granularity of use cases cannot be performed on the available driving statistics. Having to validate 1867 sets of model parameters further complicates the problem of calculation of validation mileage in SOTIF HARA.

Summary

Within this work, we have set out the goal to reduce the complexity of SOTIF HARA while still being able to define validation goals. This we started with a HARA containing 1867 lines (about 180 of those relevant for AEB functionality). We ended up with a model for calculation of the SOTIF validation range, with 32 relevant states (4 driving situations vs. 8 possible collision models), revealing almost 6-fold reduction. All the rest of the required information was obtained via driving statistics and other reference documents.

Besides the calculation of SOTIF validation targets, the Markov-based stochastic approach described above may be used for estimation of any other safety-related parameter, incl. exposure and controllability parameters.

Outlook

A simple example presented here considers a fully Markov chain with stationary transition rates, leading to considerable limitations. Based on the transition model introduced on Figure 3, semi-Markov chain may be built representing various distributions of transitions and states. Those distribution may be further parametrized, allowing the better fitting and validation of the model.

Another prospect for improvement of the approach is using better, more precise data. Here, a yearbook on traffic accident statistics from German Federal Office for Statistics was used as data source. Further sources like GIDAS study may provide for better model validation and more precise parametrization.

Contact Information

The main author:
Oleg Lurie
Safety Analyst at ZF TRW
Global Engineering Excellence - Systems Safety
TRW Automotive GmbH
Fritz-Reichle-Ring 8
D-78315 Radolfzell/Germany
Telefon/Phone +49.77732.939.1467
Telefax/Fax +49.7732.939.1820
oleg.lurie@zf.com

Definitions/Abbreviations

HARA - Hazard Analysis and Risk Assessment
AD - Autonomous Driving
ADAS - Advanced Driver Assistant System
SOTIF - Safety of the Intended Functionality
GAMAB - Performing globally at least as good as
Eq - Equation

References

1. ISO/WD PAS 21448, "Road Vehicles - Safety of the Intended Functionality," ISO Working Draft, 2013.
2. Lurie, O., "Where Lifecycle Starts and Ends," presented at *the Conference Operational Safe Vehicles for Automated Driving*, Berlin, Germany, September 19-20, 2017.
3. Koopman, P. and Wagner, M., "Challenges in Autonomous Vehicle Testing and Validation," *SAE Int. J. Trans. Safety* 4, no. 1 (2016): 15-24, doi:10.4271/2016-01-0128.
4. IEC 61882:2016, "Hazard and Operability Studies (HAZOP Studies): Application Guide," IEC Standard, Rev. March 2016.

5. Birolini, A., *Reliability Engineering: Theory and Practice* (Berlin/Heidelberg/New York: Springer, 2009).

6. VDA 702, "Situations Catalogue for E Parameters according to ISO 26262-3" (in German), VDA Standard, Rev. June 2015.

7. "Traffic. Traffic Accidents" (in German), Federal Statistical Office of Germany, Wiesbaden, 2016.

8. https://www.safercar.gov/Vehicle-Shoppers/Safety-Technology/AEB/aeb.

9. Fabris, S., "ADAS Design according to ISO 26262 and SOTIF Principles," presented at *IMECH 2017 Conference*, Birmingham, May 2017.

10. Priddy, J., Harris, A., and Fabris, S., "Method for Hazard Severity Assessment for the Case of Undemanded Deceleration," presented at *VDA Automotive System Conference*, Berlin, 2012.

Toward a Framework for Highly Automated Vehicle Safety Validation

Philip Koopman
Carnegie Mellon Univ.

Michael Wagner
Edge Case Research LLC

Validating the safety of Highly Automated Vehicles (HAVs) is a significant autonomy challenge. HAV safety validation strategies based solely on brute force on-road testing campaigns are unlikely to be viable. While simulations and exercising edge case scenarios can help reduce validation cost, those techniques alone are unlikely to provide a sufficient level of assurance for full-scale deployment without adopting a more nuanced view of validation data collection and safety analysis. Validation approaches can be improved by using higher fidelity testing to explicitly validate the assumptions and simplifications of lower fidelity testing rather than just obtaining sampled replication of lower fidelity results. Disentangling multiple testing goals can help by separating validation processes for requirements, environmental model sufficiency, autonomy correctness, autonomy robustness, and test scenario sufficiency. For autonomy approaches with implicit designs and requirements, such as machine learning training data sets, establishing observability points in the architecture can help ensure that vehicles pass the right tests for the right reason. These principles could improve both efficiency and effectiveness for demonstrating HAV safety as part of a phased validation plan that includes both a "driver test" and lifecycle monitoring as well as explicitly managing validation uncertainty.

CITATION: Koopman, P. and Wagner, M., "Toward a Framework for Highly Automated Vehicle Safety Validation," SAE Technical Paper 2018-01-1071, 2018, doi:10.4271/2018-01-1071.

Introduction

Wide-scale deployment of Highly Automated Vehicles (HAVs) seems imminent despite facing significant interdisciplinary challenges [1]. At this time, there is no generally agreed upon technical strategy for validating the safety of the non-conventional software aspects of these vehicles. Given NHTSA's "non-regulatory approach to automated vehicle technology safety" [2], it seems that many HAVs will be deployed as soon as development teams think their vehicles are ready - and then they will see how things work out on public roads. Even if pilot deployments yield acceptably low mishap rates, there is still the question of whether a limited scale deployment will accurately forecast the safety of much larger scale deployments and accompanying future software updates.

It is common to see statements to the effect that accumulating on-road miles will validate HAV system safety, especially in the context of attempting to characterize progress of development efforts. (E.g., [3], although this does not necessarily represent the actual safety approach of the company discussed.) More comprehensive discussions of the topic still tend to heavily emphasize the role of testing, even if other forms of validation are mentioned (E.g., [4, 5].). However, even with closed courses and high-fidelity simulation, there are limits to the amount of vehicle-level testing that can be done before deployment.

The scope of this paper is validation required beyond ISO 26262 compliance, with an emphasis on SAE Level 4 autonomy. Level 4 HAVs are only required to operate autonomously within a defined Operational Design Domain (ODD), which defines the specific conditions under which the system is intended to function [2, 6].

A safety validation approach for HAV autonomy that goes beyond mileage accumulation is highly desirable. Preferably, it should also be based on a falsification approach that includes concrete, testable safety goals and requirements [7]. This paper proposes a number of ways to improve HAV validation efficiency, increase effectiveness, and lead to a more defensible safety argument. A layered series of validation steps can help support a conclusion that an HAV system is acceptably safe, even in the absence of a completely specified set of traditional functional requirements for autonomy functions.

Approach

We believe that HAV validation efforts can be significantly strengthened by applying the following ideas:

1. Disentangle the disparate goals of testing by separately managing requirements validation and design validation.
2. Use higher-fidelity simulation and tests to reduce residual risks due to assumptions and gaps in lower fidelity simulations and tests.
3. Provide observability in the HAV architecture to ensure that tests are passed for the right reasons.
4. Explicitly manage uncertainty within the safety argument.

Although these ideas are based on existing practices in some domains, the novelty of HAV technology and the pace at which HAVs are being commercialized motivates a clear, unified description of how these ideas can be applied to manage and reduce the risk of aggressive HAV deployment.

Terminology

Our terminology is generally compatible with ISO 26262 [8]. The following terms are defined particularly relevant:

Risk: a combined measure of the probability and consequence of a mishap that could result in a loss event.

Safety: absence of unreasonable risk of a mishap resulting in a loss event. Level 4 HAV loss events can include fatalities potentially attributable to HAV design defects or operational faults. For initial HAV deployment, evaluation of what might constitute a "reasonable risk" will be influenced by public policy decisions.

Safety Validation: demonstrating that system-level safety requirements (safety goals) are sufficient to assure an acceptable level of safety and have been achieved.

Safety Argument (Safety Case): a written argument and evidence supporting safety validation.

Machine Learning (ML): an approach using inductive learning for system design, in which a run-time system uses the results of a learning process to perform algorithmic operations (e.g., running a deep convolutional neural network having precomputed weights). This paper assumes weights are fixed before validation. Validating dynamically adaptive ML systems that modify weights or otherwise learn at run-time is beyond the scope of this paper.

The Role of Vehicle Test and Simulation

Before describing proposed validation strategies, it is helpful to review typical uses of testing and simulation in current HAV safety assessment approaches.

Beyond ISO 26262

Dealing with many potential design and implementation defects can, and should, be done via use of an established safety standard such as ISO 26262 [8]. For areas in which even a perfectly working system might not provide completely safe functionality, an emerging standard covering Safety of the Intended Functionality (SOTIF) might be used [9]. A SOTIF standard might provide a way to deal with functions with statistically valid functionality, such as radar-based obstacle detection functions. Other issues specific to ML-based systems must also be addressed, as discussed in [10]. Overall, the problem with validating according to a V model as is typical in functional safety approaches is that ML system functionality can be opaque to humans [11]. This makes traceability problematic to the degree that humans performing traceability analysis can't analyze design artifacts [12].

Rather than attempt a design-to-test traceability approach according to the V model, we instead explore what can be done with a test-centric approach to areas beyond the obvious scope of practical application of ISO 26262 and SOTIF standards which are not designed for ML validation.

System Test/Debug/Patch as a Baseline Strategy

Historically, on-road testing has been emphasized in prototyping autonomous vehicles (E.g., [13, 14, 15, 16, 17].). The field of robotics relies heavily on "real-world" testing in order to gain an understanding of what features robots need. However, as vehicles transition from prototype to production, the approach to validation must become more comprehensive.

Basing an HAV safety argument solely on accumulating road miles is an impractical way to validate safety. Such a brute force approach takes a huge number of miles to make a credible statistical argument [18]. Beyond that, the validity of accumulated road testing evidence is potentially undermined with each software change, whether it be an update to training data, the addition of new behaviors, or just a security patch.

As a practical matter, what happens if, after billions of miles of road testing and simulation, the data shows that an HAV is not living up to its hoped-for safety goal? Will the development team (or should they) do another billion miles of road testing after fixing any observed defects? Or will the team just patch the readily reproducible bugs, test for a few miles, and declare victory, moving on to deployment? And how will the realities of the intense pressure from the race to market influence a team's interpretation of results and approach to validation?

Essentially all other industries base functional safety validation of software-based systems not on trial deployment, but rather both on testing *and* other validation approaches that can be evaluated by an independent assessor. If the HAV industry wishes to follow those precedents, it will need a way to build a methodical, defensible safety argument that can be evaluated by an independent party despite any unique validation challenges.

Limitations of Vehicle-Level Testing and Simulation

As a practical matter, it is impossible to perform enough ordinary system-level testing to assure the safety of a life-critical system. In general, this is because the exposure of an automotive fleet is so high, and life-critical safety requirements are so stringent, that testing cannot accumulate enough exposure hours to statistically prove safety [19].

For HAVs, one manifestation of the testing infeasibility problem is that unusual situations must be handled safely, but are comparatively rare in normal driving. Road testing is an inefficient way to observe rare events manifesting by chance. Closed-course testing can accelerate exposure to *known* rare events by setting them up as explicitly designed test scenarios (E.g., [20].). Evaluation might be further accelerated by skewing distributions of test cases toward the more difficult known scenarios (E.g., [21].). For example, Waymo uses both closed-course testing and extensive simulation in addition to its on-road test program [4].

Even covering known scenarios can be challenging due to resource limitations if it exclusively involves the use physical vehicles. Software-based vehicle simulation can scale up coverage of test scenarios via running simulations on many computers in parallel, but inevitably involves a tradeoff of fidelity vs. run-time cost as well as questions about completeness and accuracy of software models. Simulation suffers from the possibility of not simulating unanticipated scenarios (e.g., *unknown* safety-relevant rare events).

"Shadow mode" driving [22] and SAE Level 3 autonomy deployment [6] can increase exposure to real-world driving scenarios by monitoring a deployed fleet in which human drivers are responsible for safety. However, there is controversy as to whether a human driver can effectively supervise safety in Level 3 systems [23].

Road testing, closed course testing, simulation, and monitoring of human-tended systems all have an important place in demonstrating HAV safety. However, to be both effective and efficient they should be organized in such a way as to work together in a complementary fashion. (We recognize that many HAV developers have sophisticated but proprietary approaches to validation. In this paper we assume a naïve mileage accumulation baseline approach to illustrate the issues.)

Simulation Realism for Its Own Sake Is Inefficient

When asking why on-road testing with a real vehicle is better than simulation, a typical answer is that it is more "realistic." Ultimately testing a real vehicle in the real world is important. But realism for its own sake is an inefficient, and ultimately unaffordable, use of test resources.

The key to simulation validity is having just the right amount of realism (simulation fidelity) to get the job done. It has famously been said that *all* models are wrong, but *some* are useful [24]. Since simulations involve a model of the system, a model of the environment, and a model of system usage, it follows that no simulation is perfect.

The level of fidelity in a simulation is the degree to which it makes simplifications and assumptions about the behavior of the system. Low-fidelity simulations typically execute quickly by using simplified representations of systems (sometimes called reduced-order models), and hence in some sense are "wrong." High-fidelity simulations typically are more complex and are more expensive to execute, but contain with fewer simplifications and assumptions, and are therefore "less wrong." But both types of models can be useful.

The key to improving testing efficiency is realizing that not all realism is actually useful for all tests. As a simple example, modeling the coefficient of road surface friction is generally irrelevant to determining if a computer vision capability can see a child in the road. (The friction coefficient is likely relevant to determining if the vehicle can stop in time, but is not relevant to whether a particular geometric and environmental scenario will result in detecting a child.) This is true whether testing is done in software simulation (via modeling different road surfaces) or with a simulated test track scenario (via sand or ice on tarmac).

The key to effective and efficient simulation is considering not only the system being validated, but also the assumptions made by the various-fidelity models of the system and operational environments. Accordingly, any practical validation effort should be considered as a hierarchical series of models of varying levels of abstraction and fidelity. Viewed this way, closed-course testing is a form of simulation, because even though obstacles and vehicles involved might be real, the scenarios are "simulated." Validating HAV safety will require not only ensuring that the HAV system model is sufficiently accurate, but also validating both the environmental and usage models used to create test plans and testing simulations.

Clarifying the Goals of Testing

A robust safety validation plan must address at least the following types of defects that encompass potential faults in the system, the environment, and system usage:

- **Requirements defects:** the system is required to do the wrong thing (defect), is not required to do the right thing (gap), or has an ODD description gap.

- **Design defects:** the system fails to meet its safety requirements (e.g., due to implementation defects), or fails to respond properly to violations of the defined ODD.

- **Testing plan defects:** the test plan fails to exercise corner cases in requirements or design, or has other gaps.

- **Robustness problems:** invalid inputs or corrupted system state cause unsafe system behavior or failure (e.g., sensor noise, component faults, software defects), or an excursion beyond the ODD due to external forces.

Among the challenges faced by HAV validation are incomplete requirements and implicit representations of both requirements and design. Non-deterministic system behavior further complicates matters. These challenges will of necessity affect the approach to and goals for system testing [12] (That previous paper concentrates on identifying the challenges in validating autonomy, run-time monitoring approaches, and fail-operational approaches. We build upon that previous work here by discussing the pieces of a validation approach.).

In general, difficulties in applying traditional functional safety approaches to at least some HAV functionality motivates considering the different possible roles of testing in the overall safety validation process, as well as handling the issue of requirements incompleteness.

HAV Requirements Will Be Incomplete

A key challenge for HAV validation is that a complete set of behavioral requirements needs to be developed before behavioral correctness can be measured to provide pass/fail criteria for testing. For example, while efforts are underway to document vehicle behaviors and scenarios (e.g., the Pegasus Project [25]), there is not a complete, public set of machine-interpretable of traffic laws that includes exception handling rules (e.g., when and how exactly can a vehicle cross a center dividing line, if present, to avoid a lane obstruction?). We use the term "requirements" in this paper primarily to refer to system-level behavioral requirements, although the concepts can apply in other ways as well.

Requirements gaps are a primary motivation for on-road vehicle data-gathering operations, which sometimes are loosely referred to as "vehicle testing." The general strategy of inferring system requirements from road test data also affects the completeness of test plans, in that there will be testing gaps corresponding to gaps in system behavioral requirements (e.g., unknown and therefore missing behavioral scenarios).

It is important to note that, strictly speaking, systems that use on-road data as the basis for training machine learning do not ever identify requirements per se. Rather, the training data set is a proxy for something akin to requirements [12]. In other cases, analysis of on-road data might be used to construct some level of explicitly stated requirements. Successfully validating an HAV requires that test plans capture and exercise the required behaviors, even if expressed implicitly. Regardless of the form, these requirements or proxies for requirements are likely to be incomplete for many initial HAVs deployments.

Vehicle Testing for Debugging Can Be Ineffective

A common view of system-level testing is that it is a way to discover software defects ("bugs") and remove them. However, there is a steep diminishing returns problem for vehicle-level testing. Once the easy bugs have been found that involve typical driving scenarios, it can get dramatically more difficult to find additional defects. This is especially true for defects that require very precisely specified initial conditions, involve timing race conditions, or involve recovery from computational run-time faults that are difficult to induce using ordinary vehicle interfaces. This problem is even worse in robotics, in which we have observed that minute variations in lighting and geometry can trigger unreproducible bugs. In general, it can be expected that many such subtle bugs will escape detection and diagnosis during any reasonable amount of vehicle testing, and will be non-reproducible for practical purposes. However, they will surely show up in the field in high exposure applications such as automotive systems.

Beyond an efficiency problem, any project that uses vehicle testing as its primary mechanism of defect removal has a fundamental problem in its safety world-view. Testing can prove the presence of bugs but not their absence [26]. Moreover, when all the bugs found by test have been fixed, the bugs that are left are ones that the testing procedures are not designed to find (the Pesticide Paradox [27]). Thus, even if vehicle-level testing finds no problems at all, that does not mean the vehicle's software is necessarily safe. This line of reasoning is simply another path to concluding that vehicle-level testing alone is an untenable approach to proving system safety.

Vehicle Testing as Requirements Discovery

Some forms of "vehicle testing" are actually aimed at requirements discovery. Examples of areas in which still-maturing HAV development efforts might well have requirements gaps include:

- Detecting and evading novel road hazards

- Handling of exceptional situations that require violating normal traffic rules

- Unusual vehicle configurations, surfaces, and paint jobs

- Misleading but well-formed map data

- Novel road signs and traffic management mechanisms specific to a micro-location or event

- Unusual road markings and vandalism

- Emergent traffic effects due to HAV behaviors

- Malicious vehicle behavior (humans; compromised HAVs)

While HAV designers should design for known requirements, continual novel operational "surprises" are inevitable in the real world for the foreseeable future. A primary rationale for Level 4 automation rather than full Level 5 autonomy is so that the HAV does not have to handle all possible scenarios. Rather, a significant feasibility benefit of Level 4 autonomy is that it is permitted to exhibit a graceful failure when outside its ODD so long as its failure response is safe. Indeed, it would be no surprise if Level 5 autonomy remains an elusive goal over the long term, with Level 4 autonomy asymptotically approaching - but never actually attaining – complete automation in all possible operating conditions and scenarios.

It is important to point out that Level 4 autonomy does not relieve an HAV safety assurance argument from having to deal with all possible scenarios, including ODD violations and novel scenarios. The general concept of an ODD seems to assume that one of the following two situations must be true: (1) there is some external guarantee that the HAV won't encounter a situation it can't handle well due to a highly reliable

ODD constraint (e.g., robustly predicting kangaroo road hazard behavior [28] is generally not required on North America public roadways), and/or (2) the HAV will reliably detect that it is in a situation outside its ODD and bring the vehicle to a safe state (e.g., a vehicle not rated for kangaroo road hazards might be geo-fenced out of a wild animal park and the continent of Australia). In reality, it is possible that the ODD will be violated without being detected due to gaps in understanding the full scope of an ODD (e.g., the designers never considered kangaroos in the first place), or gaps in the validation plan that omit testing relevant ODD constraints.

An appropriate use of on-road operation is finding requirements gaps. Encountering some unexpected scenarios will result in a requirements update, while others result in a modification either of ODD parameters or ODD violation detection requirements. It is important that the HAV be acceptably safe when it first encounters such an ODD "surprise." Accomplishing this is problematic since, by definition, such a scenario is unexpected and therefore not a designed part of any test plan.

Since no validation approach is perfect, it is likely that some design defects will escape and be found via road tests, or even in deployed vehicles. However, this should be a very small fraction of the total number of defects found in the system, and those defects should result in safe behavior even if that behavior does result in a system safety shutdown or other loss of availability. If an excessive fraction of defects escape detection during the development cycle and aren't seen until road testing, that is indicative of a systemic problem with requirements, test plan, or some other element of the validation approach. As with any safety critical design process, defect escapes to production systems should be cause for a significant response to correct any safety process problems that contributed to the situation.

Separating Requirements Discovery and Design Testing

A crucial perspective regarding the role of on-road testing is that accumulating miles in a search for missing requirements isn't really "vehicle testing" in the traditional sense at all. It is a requirements-gathering and validation exercise. On the other hand, whether on-road data or some combination of simulation, synthesized data, and recorded data are the primary means for testing a particular HAV design is more at the discretion of the design team. So long as the design is validated according to an adequately complete set of requirements, on-road testing need not (and in practice should not) be the only testing performed.

Thus, one way to reduce the time and expense of HAV validation is to separate (1) on-road for requirements gathering from (2) design and implementation validation. There is no obvious way around needing billions of miles of on-road experience to seek out rare but dangerous events that need to be mitigated by system safety requirements. But that doesn't mean that design validation needs to re-do those billions of miles for every design change – at least if a more sophisticated approach is taken beyond brute force system-level testing.

Vehicle Testing to Mitigate Residual Risks

We can generalize upon the notion that on-road testing should primarily emphasize requirements validation, while lower level simulation and testing should emphasize the validation of design and implementation. In general, any level of simulation (including "simulated" aspects of vehicle testing) has a particular level of fidelity as previously discussed. That means that it is also "wrong" – just as all models are wrong – in some aspect due to its simplifications and assumptions.

Improving testing efficiency can be accomplished by focusing the test plans for each level of fidelity on checking the assumptions and simplifications of lower-fidelity levels of simulation. At the same time, pushing as much simulation as possible to lowest practical level of fidelity will decrease simulation costs. For example, simple coding defects should be found in subsystem simulation (or even pre-simulation via traditional software unit test and peer reviews). On the other hand, rare event requirements gaps might be best found in on-road testing if they are due to unforeseeable factors. This leads to an approach based on mitigating residual risks for each level of simulation fidelity, as discussed in the following section.

A Layered Residual Risk Approach

Since complete human-interpretable design and requirements information is unlikely to be available for HAVs in the near term, some approach other than, or in addition to, the traditional V model must be used for validation. To do this, we need to start with at least a (possibly incomplete) set of safety requirements. Then, we must find a way to trace some combination of road testing, closed course testing, and simulation results back to those safety requirements.

Validation According to Safety Requirements

At the highest level, we need some type of system requirements to be able to determine whether tests actually pass or fail. If functional requirements are not fully spelled out, then we need something else. The good news is that optimal performance may not be needed to provide safety. Rather, simpler requirements are likely to be sufficient to define safe operation.

For example, we have found that a list of unsafe behaviors that are forbidden based on safety envelopes can be sufficient for some autonomous vehicle behaviors [29]. In that case, testing can be traced to explicitly stated safety requirements even if the functional requirements themselves are opaque or undocumented. One way to specify safety envelopes is using runtime invariants allocated to a distinct safety checker functional block [30]. As a simple example, a safety envelope for lane-keeping could be that the vehicle stays within its lane boundaries plus some safety margin. This is much simpler to specify and use as a test success oracle than checking perfect implementation of a complex algorithm that optimizes the vehicle's lane position according to road geometry and traffic.

While tracing tests to stated safety requirements can be helpful, we have found via experience that too often safety requirements are poorly understood, or not even written down at a useful level of detail. While a vague notion that mishaps should not occur is a starting point, there must also be a concrete and specific way to determine if a test has shown that a system is safe or not. In practice, we have found that a set of partial runtime invariants that specifies a combination of safe and unsafe system state space envelopes can be evolved over time in a continuous improvement approach in response to the results of testing and simulation. In other words, one way to approach the problem of missing safety requirements is to start with simple set of rules and elaborate them over time in response to tests that violate those simplistic rules. False positive and false negative rule violations can drive refinement of the rule set. Generally, this evolution works best if it starts with an under-approximation of the safe operating envelope (increasing the high false positive rate) and progressively adds additional envelope area (and accompanying test oracle detail) when analysis shows that doing so is a safe way to increase envelope permissiveness.

If an HAV design team attempts to determine safety requirements via machine learning-based approaches, it will be important for them to express the results in a way that is interpretable to human safety argument reviewers. However, it is unclear how that might be done. At this point we recommend using more traditional engineering approaches to defining safety requirements to avoid the same problem of inscrutability that befalls ML-based functionality.

Basing Validation on Residual Risks

While a safety envelope approach can simplify the complexity of creating a model of requirements to use for pass/fail criteria, HAV testing will still need to run a huge number of scenarios to attain reasonable coverage. Ideally as much as possible will be done with comparatively inexpensive, low-fidelity simulations. Then the approach should add fidelity not just for the sake of undifferentiated "realism," but rather for the sake of *reducing the residual risks* due to simplifications made by low fidelity simulations.

Managing Residual Risks

The important relationship between high- and low-fidelity simulation runs should not be one of "sanity checking" or statistical sampling, but rather one of emphasizing validating the correctness of assumptions and simplifications made at lower fidelity levels. In other words, for each aspect in which a particular level of fidelity model is "wrong" in some respect, a higher fidelity simulation (including potentially various types of physical vehicle testing) should assume the burden of mitigating that residual safety validation risk.

This approach is different than the usual notion of model validation in an important way. Higher fidelity levels of simulation are not only used to validate the correctness of lower fidelity models, but must also be explicitly designed to emphasize checks of the assumptions and simplifications that are known to be present as simulations are run. A primary goal of a higher fidelity model should be to mitigate that residual risk by not only checking the accuracy of lower fidelity simulation results, but also by checking whether assumptions made by lower fidelity models are violated when the higher fidelity simulation is performed. As a simple example, if a simplified model assumes 80% of radar pulses detect a target, a higher fidelity model or vehicle test should flag a fault if only 75% of pulses detect a target - *even if* the vehicle happens to perform safely according to the higher fidelity model. The assumption of 80% detection rates is a residual risk of the lower fidelity simulation that makes that assumption. Violating that assumption invalidates the safety argument, even if a particular test scenario happens to get lucky and avoid a mishap.

This approach fundamentally affects the design of a simulation and test campaign. For example, consider a simulation that explores obstacle placements across the field of view. The simulation arranges obstacles in the environment with very precise resolution, but uses only crude stick-figure simulated pedestrian objects in static positions at a fixed orientation. Doing thousands of additional high-fidelity vehicle tests while varying obstacle placement would be expected to yield a low marginal validation benefit over exhaustive simulation results, especially if the simulation exercises the actual geometry processing code that will be deployed in the HAV. That is because in this example obstacle placement relative to the vehicle is not the primary source of residual risk after simulations are completed. The main residual risk revolves around the pedestrians. The low-fidelity simulation assumes stick figure people, thereby omitting consideration of people carrying large objects, people wearing clothing that significantly distorts sensor signals, different rotational positions with regard to vehicle sensors, and so on.

By the same token, any improvement of simulation capability should not merely strive to make the simulation higher fidelity in every possible dimension. For example, modeling road obstacle placement down to the nanometer rather than the millimeter is not likely to be a generally productive use of simulation resources. Rather, simulation fidelity improvements should be made to replace required system level tests with simulations (e.g., adding surface texture capability as well as a wider variety of geometrical shapes and orientations for the previous stick figure example).

This does not mean that simulation model verification and validation (e.g., as described in [31]) should be neglected. Rather, the point is that even a perfectly validated model at a particular level of abstraction leaves residual risk. Part of the risk is because of the possibility of an incomplete testing campaign, which amounts to not fully mitigating risks inherited from lower fidelity simulation or not fully covering the areas assigned to the level of fidelity in question. Another part of the risk is due to safety considerations that have been intentionally excluded at a particular level of abstraction, which corresponds to risks passed up the line to the next higher level of fidelity.

Thus, the time-honored approach of using runs of varied simulation fidelity [32] still makes sense for HAVs. The art is in making sure that simplifications in lower fidelity tests are explicitly managed and mitigated as validation risks.

The approach of accelerated evaluation via biasing tests towards difficult scenarios [21] is complementary to a residual risk approach. Emphasizing difficult scenarios is intended to winnow redundant nominal path tests from the test set while still covering off-nominal behaviors, edge cases, and complex environmental interactions. On the other hand, residual risk mitigation addresses the potential problem of risks due to simplifications and unchecked assumptions made by lower fidelity layers of a simulation and testing plan.

An Example of Residual Risks

Table 1 shows a simplified example of residual risks that should be considered with an HAV testing and simulation plan. The residual risks at the top of the table tend toward requirements gaps (unexpected scenarios and unexpected environmental conditions). In comparison, the other residual risks tend toward a combination of simplifications

TABLE 1 Hypothetical validation activities and threats to validity

Validation Activity	Residual Risks (Threats to Validity)
Pre-deployment road tests	unexpected scenarios, environment
Closed course testing	*As above, plus:* Unexpected human driver behavior, degraded infrastructure, road hazards
Full vehicle & environment simulation	*As above, plus:* simulation inaccuracies, simulation simplifications (e.g., road friction, sensor noise, actuator noise)
Simplified vehicle & environment simulation	*As above, plus:* inaccurate vehicle dynamics, simplified sensor data quality (texture, reflection, shadows), simplified actuator effects (control loop time constants)
Subsystem simulation	*As above, plus:* subsystem interactions

driven by speed/fidelity simulation tradeoffs (e.g., sensor data quality) at the mid-level, and potential design issues (e.g., subsystem interactions) at the lowest level.

Revisiting the previous obstacle detection example, this means that higher fidelity levels such as physical vehicle testing should not primarily focus on different sizes and placement of obstacles. Rather, they should focus on things such as dirt on objects and sensors, and other aspects that might not be handled by software-only simulation tools. In other words, vehicle testing should mostly concentrate not on reproducing simulation results, but rather on challenging any known weak points of the simulation methodology. Specifics will vary. The point is that all simulation tools have limitations of some sort that require further validation efforts.

For the example shown in Table 1, closed course testing should not focus on unexpected human driver behavior, degraded infrastructure, or road hazards, because mitigating those threats is the primary reason to do pre-deployment road tests. Expected behaviors, road hazards, and so on should be handled with testing and simulation. It is unexpected problems that can't be addressed, because an unexpected problem is by definition not something that can be explicitly included in a test plan.

It is important to avoid burdening higher level system testing with addressing risks that should properly be dealt with at lower levels. Continuing the example, closed-course testing should not be significantly concerned with normal vehicle dynamics, and ordinary issues of sensor data quality and actuator effects, since those can be taken care of with software-based simulation. Vehicle testing should also not be used to brute force test obstacle placement and geometries that can more be dealt with in a more cost-efficient way with simplified vehicle and environment simulation that exercises just the vehicle's obstacle-handling code. Prototyping tests with a real vehicle on a closed course might make sense when validating the simulation capability. But executing the actual vehicle testing campaign should be done at the lowest practical level of simulation fidelity for each aspect of the test plan as much as possible to reduce time and costs.

The overarching idea is that the *primary* emphasis in each level of validation should be on residual risks inherited from the next lower level, especially when re-running existing simulation test suites on a system that has been modified so as to ensure that the system is still safe. Extensive sampling to exhaustively replicate the results of lower fidelity simulation and testing is wasteful at best, and at worst gives a false sense of security if the random sampling does not cover residual risks.

Improving Observability

Given a thorough simulation- and vehicle-based test plan, sufficient controllability and observability must be provided to yield a credible safety validation outcome.

Controllability and Observability

Controllability is the ability of a tester to control the initial state and the workload executed by a system under test. *Observability* is the ability of the tester to observe the state of the system to determine whether a test passed or failed [33].

Controlling test scenarios to elicit a particular autonomous system behavior is difficult [12]. This is due to a combination of the use of stochastic methods (e.g., randomized path planners), sensitivity to initial conditions (e.g., exactly repeatable sensor alignment within a test environment), variability in actuator outputs (e.g., unexpected variations in environmental interactions with actuators), and computational timing variations.

A useful approach to improving controllability is to use simulation that can avoid physical world randomness and constraints. Beyond that, a system testing interface can be provided that forces the system into an initial state for testing. For example, a path planner might be tested in a repeatable manner if its internal pseudo-random number generator can be set to a predetermined seed value. As a practical matter, deterministic testing requires that the HAV software be intentionally designed to provide a deterministic testing capability. It can be difficult to mitigate sources of non-determinism in software after it has been constructed.

Observability can be a more difficult problem. For example, in a vehicle-level obstacle test the vehicle either leaves sufficient clearance as it passes an obstacle or it does not. But, even if the system "passes" a test by not colliding, that could simply be due to the system getting lucky in avoiding an obstacle it did not even know was there. The system might hit the obstacle on the next test run - or perhaps hit it 2000 test runs later. This lack of observability is one facet of the robot legibility problem, which recognizes the difficulty of humans understanding the design, operation, and "intent" of a robotic system [34] (The additional role of legibility in HAV interaction with human drivers is an important one, but beyond the scope of this paper.).

While one can argue that it is unlikely a system will repeatedly pass tests by dumb luck, the sheer number of test parameters involved makes the "repeatedly" part of that argument expensive. And, regardless of how many tests are run, it is difficult to achieve an extreme level of statistical significance via testing for life-critical assurance levels. (Even a 99.99% confidence level for a system avoiding a detected child in a crosswalk seems problematic if it could result in one out of 10,000 children being hit.) Thus, there will always be a residual risk that some combinations of scenario elements pass tests repeatedly due to a lucky streak rather than due to a safe design.

Software Test Points

Rather than relying only upon system-level behavior and brute force repetition to determine if a test passes, a more efficient testing approach can be to insert software test points into the system to improve observability. For example, if sensor fusion dependability is a residual risk due to simulation limitations, a relevant test point for closed course vehicle simulation would be monitoring the computed certainty level of a sensor fusion results. That would provide information about whether a test obstacle is being avoided with the intended margin of error rather than by luck. (The issue of software test points potentially disturbing the system under test can be resolved by architecting test points in as a permanent part of the system. This will in turn facilitate data collection in the deployed system.)

Software test points also facilitate monitoring for safety argument assumption violations during fleet deployment. The previously discussed 80% detection rate assumption example can be monitored not only during testing, but also during full scale vehicle deployment to detect assumption violation escapes into fielded systems.

Passing Tests for the Right Reason

When a human takes a driver test, the test examiner has a fairly accurate (or at least useful) mental model of the driver behind the wheel. If the driver changes lanes without making eye contact with a rear-view mirror or otherwise checking for vehicles in the destination lane, the examiner knows that the driver got lucky in executing a collision-free lane change instead of behaving properly. With an HAV, this type of assessment is more difficult, because it is unclear what the "tells" are for a machine exhibiting safe behavior vs. getting lucky with unsafe behavior. That is especially true if requirements and design are not traceable via a V-based safety process.

If HAV safety is to be based in part on a driving-test type event, then the examiner must know that the HAV not only behaves the right way, but also behaves the right way for the right reason. Even without a formal driver test, being able to reasonably infer causality of actions from explicit system information can reduce testing costs compared to a brute force statistical approach. Having an HAV self-report regions of saliency [35], bounding boxes on objects, and so on is not a new idea. However, explicitly including such capabilities in a safety argument can reduce testing cost if exploited in the right way. This may motivate further work to verify that self-reporting and explainability mechanisms work reliably.

One way to couple scenarios with behaviors is to have the HAV self-report the scenario it thinks it is in, or the various scenario elements that it thinks are in play. As an example, rather than just performing a vehicle lane change when it can, the vehicle might report: "I want to change lanes … I am checking the next lane and there is a car there but it is sufficiently far behind me that I am clear … I am starting to change lanes … I am continuing to monitor that the lane is still clear … the car behind me is speeding up to close the gap …" and so on. Some HAV architectures might provide this level of observability already. The question is how formally such information is used by the validation strategy. Moreover, many popular approaches (e.g., end-to-end deep learning) explicitly eschew architectural modularity, which tends to degrade observability. They do so with the goal of achieving higher performance, a tighter implementation, and less development effort [36]. Lack of observability has the potential to exact a high price in terms of validation effort or deployment risk for such systems.

An effective driving test should require not only correct behavior, but also a correct introspective narrative of why the HAV is acting the way it is. That is a good start, but we must then must question the integrity of a machine's explanation for its actions. However, we argue that deciding whether to trust an explicit explanation is an easier to solve problem than having to infer (and then trust) an opaque implicit explanation via behavioral observa-

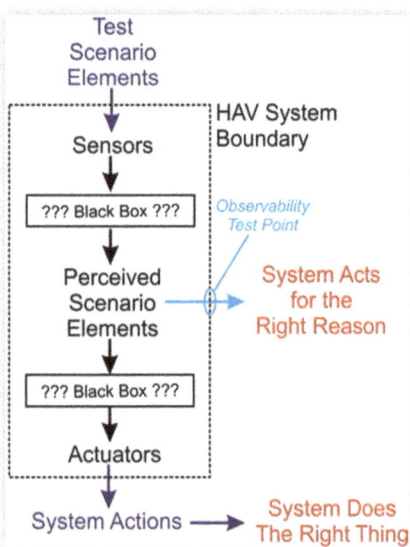

FIGURE 1 System validation should determine that the system does the right thing for the right reason.

tion. Either way, a decision must be made about whether the vehicle will do the right thing in future circumstances that are not exact matches to training and test data sets. The advantage of an explicit explanation is that the validity of that mechanism can be made falsifiable if it is required to match the test plan narrative. In designing safety-critical systems, we prefer explicit, verifiable, simple patterns that might be less performant over those that are highly-optimized but opaque. We have reason to believe this trend will hold for HAVs when considering the consequences of attempting to deploy difficult-to-validate systems.

Architecting such a system will require introducing or identifying observability for the purpose of validation. This might be accomplished by having a tool that converts existing data to human-interpretable form, adding a test point to the system architecture, or re-architecting the system to intentionally create new forms of human-interpretable data (Figure 1).

For machine learning systems, this approach suggests a somewhat unusual design strategy. Rather than having an ML system learn its own feature set for achieving an outcome, it must meet two concurrent goals: (1) display the right behavior, and (2) display a set of narrative descriptions or other explanation that matches its behavior. One way to accomplish this is to use models of environments and usage scenarios to define the set of ML outputs that must be learned. While this might be seen as additional design burden and overhead, such might be the price for being able to know whether a vehicle is actually safe enough to deploy.

To avoid a mismatch between behavior and narrative, one possibility is to arrange the ML system so that it operates in two disjoint phases: first creating the narrative, and then using the narrative as inputs for its behavior, as shown in Figure 1. The first phase might build on existing work on creating descriptions of scenarios and hierarchical classification (E.g., [37, 38].). The system actuation should be responsive to the narrative by having the second stage be fully dependent upon the outputs of the first stage. This dependency mitigates the risk of a parallel narrative construction being generated that does not actually match the system's actuation strategy.

Coping with Uncertainty

Knowns and Unknowns

Even with a validated and apparently defect-free system, there is still residual risk from problems due to incomplete understanding of the system and its requirements. These include at least the following potential types of issues:

- Emergent system properties and interactions that are not accounted for at the appropriate validation phase

- Unexpected correlated faults in areas for which safety depends upon implicit independence assumptions

- Scenario and environment exceptions that happen too infrequently to be diagnosed by pre-deployment road tests

- Uncertainty as to the arrival rates of unmitigated hazards that were assumed to be extremely infrequent

- In-range system inputs that activate unexpected defects in ML-based components

There are doubtless other types of defects that are not listed above and are not included in at least some HAV validation plans. Those are the famous "unknown unknowns" [39] that can compromise safety and cause other system failures.

Dealing with Unknown Defects

While approaches such as safety envelopes can help, in the end, there is no way to completely mitigate residual risks from unknown types of defects. However, the arrival of unexpected faults can be monitored to increase confidence over time that the residual risk is sufficiently low. It is essential to recognize unknown problems as a residual risk that must be monitored and mitigated as necessary throughout the life of the fleet. A confidence assessment framework [40] that has been extended to include unknown unknowns is one approach that could provide a way to manage residual risks.

Each time a surprise causes a safety problem, additional steps should be taken to address underlying system and safety argument assumptions that are invalidated by the newly discovered issue (this is in accordance with existing safety practices, e.g., [8]). It is important to do a root-cause analysis of unexpected faults to at least determine if a problem is a known unknown (in which case now you know more about it), or an unknown unknown (in which case you need to add a category of defect type to your validation plan and safety argument to address this new unexpected source of problems).

HAV Maturity

There is substantial intuitive appeal to having a "driving test" as part of HAV validation. However, the analogy of taking an HAV out for a road test similar to a human driving test falls short because there are actually two key elements to a human driving test. The first element is the obvious, overt requirement that the driver must show basic driving knowledge and proficiency, including a driving skills test.

The second and more subtle part of passing a driving exam is that the driver must be approximately 16 years old, depending upon locale. That age requirement serves as a proxy for having reasonably mature judgment that can handle exceptional situations and generally behave in a reasonable manner when encountering a novel unstructured situation. In the real world, correct vehicle operation depends in part upon traffic regulations. However, it also depends upon whether a police officer expertly, though subjectively, thinks the driver behaved in a reasonable and responsible manner for a given situation. ("Plays well with others" is an important HAV characteristic, especially in mixed human/HAV traffic.)

While it is possible (some say certain) that HAV behavior *can* be safer than a person given human frailties, how to measure HAV "maturity" to ensure that this desirable outcome is fully achieved remains an open question.

One way to measure HAV maturity is to deploy vehicles and see how they do. That is one of the arguments for deploying SAE Level 3 automation, which in effect uses a human in the role of an adult supervisor who monitors the junior driver during learner's permit operation. However, there are legitimate concerns that driver supervision will be ineffective over long periods of exposure due to driver dropout, especially when automation fails infrequently [41].

We propose two different approaches for evaluating HAV maturity beyond developer adherence traditional safety critical software engineering principles. The first way is ensuring that the HAV passes a detailed technical driving skill test *for the right reasons*, and the second way is monitoring whether the HAV *validation assumptions and residual risk monitoring* hold up when it is deployed in the real world. In other words, the system design might be considered to be mature if the vehicle can explain its behavior in a way that makes sense to a human and its safety case assumptions hold true in operation.

HAV Probation: Monitoring Assumptions

Any responsible decision to deploy an HAV must be more sophisticated than simply saying "we fixed all the bugs we found so we must be perfect," because that is never a reflection of reality. There is always one more bug [42]. Rather, a safety argument based on phased validation should at least be made based on measuring rates of defect escapes from each phase of validation. This argues that observability test points should be retained and monitored all the way through to fleet deployment. Doing so permits monitoring system design maturity by ensuring that there are no vehicle operational situations that invalidate assumptions. If a high rate of assumption violations is detected by runtime monitoring, that can provide valuable feedback to the design team of an impaired safety margin. In this manner, issues with the safety argument can be identified even if no actual mishaps have occurred.

As another example beyond the previously discussed assumption violation example, consider the somewhat controversial topic of disengagement reports for HAV road testing [23]. Clearly, not all disengagements are created equal, especially given that various teams are likely to have different false positive rates for triggering disengagements.

Using an approach such as Orthogonal Defect Classification (ODC) [43] might reveal, for example, that some disengagements are due to problems that should have been caught in subsystem simulation, while other disengagements are due to the discovery of a require-ment or scenario gap at the highest level. While one expects that HAV development teams do some sort of analysis on disengagements, a methodical analysis that maps defects back to residual risks identified in a validation plan has significant potential benefits, such as providing a health indication for the safety argument and the HAV's overall maturity level.

This approach can support an external assessment of autonomy validation by presenting a well-reasoned set of risk mitigation goals for each phase of validation. Those can be paired with data on defect escapes as measured by relevant observability points during simulation, vehicle testing, and deployment. All this implies that the "driver test" is not actually a one-time event, but rather involves a continual "license" renewal process based on collecting and analyzing field data on defect escapes over the life of the system.

Deploying with Residual Risks

It is important to acknowledge that this discussion has contemplated fielding HAVs that have residual risks, and in particular, potential gaps in requirements and design verifica-tion. This is inherent to the domain and the technology being deployed. It will be some time before statistically defensible amounts of data are accumulated to argue that the residual risks fall below the usual safety critical system safety thresholds (for example, below one catastrophic vehicle mishap per 10^9 or 10^{10} operational hours). Given the current HAV market and regulatory climate, it seems likely that public deployment will scale up before such data is collected.

Regardless of the appeal of fielding HAVs, is essential that the deployment be done in a responsible manner. In particular, residual risks should not be accepted blindly. Rather, residual validation risks at all levels should be explicitly understood as well as monitored during deployment. As an example, credible arguments that a particular category of residual risk is likely to result in low consequence, highly survivable, or extremely infrequent mishaps might be a legitimate motivation to determine they are "acceptable" even if the full extent of the risk is unclear. However, any such argument should be supported by monitoring field feedback data to determine if the assumptions that support the acceptance of such risks are actually true, preferably without waiting for an accumulation of serious loss events.

Ultimately ethical issues arise, such as whether it is better to deploy imperfect technology if there is an expected net savings of life [44]. Safety professionals in particular face a pragmatic choice as to whether they participate in a release of a safety-critical system with unknown (and unknowable, in the short term) but safety risks, or they miss an opportunity to improve the relative safety of HAVs that are bound to be deployed with or without their help. A goal of this paper is to provide a framework for validating such systems before they are deployed that will improve the developers' ability to identify and manage accepted risks.

Conclusions

Summarizing, we describe an approach to HAV validation that includes the following elements:

- A phased simulation and testing approach that emphasizes testing to mitigate residual validation risks from the previous phase while exploiting the speed vs. fidelity scalability properties inherent in testing and simulation.

- Observability points to produce human-interpretable data that both detect defect escapes from lower fidelity simulation phases and demonstrate the system is doing the right thing for the right reason.

- Explicit differentiation of the various roles of testing from checking for requirements gaps to checking for design faults, and matching each type of testing with a relevant portion of a phased validation approach.

- A run-time monitoring approach to managing identified risks, catching assumption violations and unknown unknowns as they arise in fielded systems.

This approach can be expected improve validation effectiveness compared to a brute-force testing campaign because it explicitly links testing and simulation activities to the risks being mitigated. This in turn permits concentrating effort on the sweet spot of defect detection for each particular level of simulation and test fidelity. The approach can also be expected to improve testing efficiency by concentrating each phase of testing on mitigating risks inherited from the preceding phase, without wasting resources revisiting low-risk conclusions or attempting to address out-of-scope risks that belong to other testing phases. (Other forms of validation beyond testing are also important, such as employing ISO 26262 approaches to appropriate portions of system functionality.)

We recognize that, due to the challenges of conclusively establishing the safety of machine-learning functions, the approach presented here will yield an ongoing process of iterative improvement rather than air-tight proofs of safety. However, the approach will serve to underscore where assumptions are being made, and where safety case evidence is missing. One way of validating the approach as well as the system is to create a Goal Structuring Notation-organized safety case (e.g., starting with [45]) and including explicitly stated assumptions to complete the argument. Each assumption identifies the residual risks for a testing or simulation technique. Assumptions that are checked by other validation approaches form part of the safety argument chain. Assumptions that can't be validated at design time are residual risks that are especially important candidates for run-time monitoring in deployed systems.

At some point, designers will have to decide on a responsible deployment plan that might involve taking risks that are judged to be acceptable according some defensible set of technical and social criteria. To minimize unmitigated residual risks, we suggest avoiding architectures in which autonomy that can't be validated using traditional safety approaches is the sole means of ensuring operational safety. One alternative is using a safety checker that can be rated appropriately according to ISO26262, such as a safety envelope monitor [46].

While it is always better to ensure that all residual risks are known and mitigated to an acceptable level, it is clear that HAVs are going to be deployed even if there are places in which the safety argument contains risks that are not completely understood. The approach discussed in this paper provides a framework for establishing an initial safety argument based on multiple levels of simulation and testing fidelity. It also provides hooks for continuous improvement based on monitoring assumption violations and other residual validation risks during the course of testing and deployment.

Our next steps are refining techniques for establishing traceability from safety requirements to test and simulation plans, and applying this approach to at-scale validation activities.

Contact Information

Dr. Philip Koopman is an Associate Professor of Electrical and Computer Engineering at Carnegie Mellon University, where he specializes in software safety and dependable system design. He also has affiliations with the Carnegie Mellon University Robotics Institute, National Robotics Engineering Center (NREC) and the Institute for Software Research. He is CTO and co-founder of Edge Case Research, LLC. koopman@cmu.edu

Michael Wagner is CEO and co-founder of Edge Case Research, LLC, which specializes in software robustness testing and high-quality software for autonomous vehicles, robots, and embedded systems. He is also affiliated with the National Robotics Engineering Center. mwagner@edge-case-research.com

Definitions/Abbreviations

HAV - Highly Automated Vehicle
ODD - Operational Design Domain
NHTSA - National Highway Traffic Safety Administration
SOTIF - Safety Of The Intended Functionality
V model - A software development model that includes requirements and design on the left side of a "V" with verification and validation on the right side of the "V"

References

1. Koopman, P. and Wagner, M., "Autonomous Vehicle Safety: An Interdisciplinary Challenge," *IEEE Intelligent Transportation Systems Magazine* 9, no. 1 (Spring 2017): 90-96.

2. NHTSA, "Automated Driving Systems: A Vision for Safety," US Department of Transportation, DOT HS 812 442, September 2017.

3. Carson, B., "Uber's Self-Driving Cars Hit 2 Million Miles as Program Regains Momentum," *Forbes* (December 22, 2017).

4. Waymo, "On the Road to Fully Self-Driving: Waymo Safety Report," 2017, https://goo.gl/7HUiew.

5. General Motors, "2018 Self-Driving Safety Report," 2018, https://goo.gl/ruLJvV.

6. SAE, "Automated Driving (from SAE J3016)," accessed 10/13/2017, http://www.sae.org/misc/pdfs/automated_driving.pdf.

7. Wagner and Koopman, A Philosophy for Developing Trust in Self-Driving Cars, Meyer, G. and Beiker, S., editors, *Road Vehicle Automation 2, Lecture Notes in Mobility* (Springer, 2015), 163-170.

8. "Road Vehicles—Functional Safety—Management of Functional Safety," ISO 26262, 2011.

9. "Road Vehicles—Safety of the Intended Functionality," ISO/ WD PAS 21448. Under development.

10. Salay, R., Queioz, R., and Czarnecki, K., "An Analysis of ISO 26262: Using Machine Learning Safely in Automotive Software," https://arxiv.org/pdf/1709.02435.pdf.

11. Dosovitskiy, A., and Brox, T., "Inverting Convolutional Networks with Convolutional Networks," *CoRR*, abs/1506.02753 (2015).

12. Koopman, P. and Wagner, M., "Challenges in Autonomous Vehicle Testing and Validation," *SAE Int. J. Trans. Safety* 4, no. 1 (2016): 15-24, doi:10.4271/2016-01-0128.

13. Urmson, C. et al., "Autonomous Driving in Urban Environments: Boss and the Urban Challenge," *Journal of Field Robotics* (2008): 425-466, doi:10.1002/rob.

14. Levinson et al., "Towards Fully Autonomous Driving: Systems and Algorithms," *IEEE Intelligent Vehicles Symp.* (June 5-9, 2011):163-168.

15. Broggi et al., "Extensive Tests of Autonomous Driving Technologies," *IEEE Trans. Intelligent Transportation Systems* 14, no. 3 (September 2013): 1403-1415.

16. Ziegler, J. et al., "Making Bertha Drive—An Autonomous Journey on a Historic Route," *IEEE Intelligent Transportation Systems Magazine, Summer* (2014): 8-20.

17. Aeberhard, M. et al., "Experience, Results and Lessons Learned from Automated Driving on Germany's Highways," *IEEE Intelligent Transportation Systems Magazine, Spring* (2015): 42-57.

18. Kalra, N. and Paddock, S., "Driving to Safety: How Many Miles of Driving Would It Take to Demonstrate Autonomous Vehicle Reliability?" Rand Corporation, RR-1479-RC, 2016.

19. Butler and Finelli, "The Infeasibility of Experimental Quantification of Life-Critical Software Reliability," *IEEE Trans. SW Engr.* 19, no. 1 (January 1993): 3-12.

20. Madrigal, A., "Inside Waymo's Secret World for Training Selfdriving Cars," *The Atlantic* (August 23, 2017).

21. Ding, Z., "Accelerated Evaluation of Automated Vehicles," accessed 10/15/2017, http://www-personal.umich.edu/~zhaoding/accelerated-evaluation.html.

22. Golson, J., "Tesla's New Autopilot Will Run in "Shadow Mode" to Prove that It's Safer than Human Driving," *The Verge* (October 19, 2016).

23. Davies, A., "The Very Human Problem Blocking the Path to Self-Driving Cars," *Wired* (January 1, 2017).

24. Box, G., "Robustness in the Strategy of Scientific Model Building," MRC Technical Summary Report #1954, University of Wisconsin, Madison, 1979.

25. Putz, A., Zlocki, A., Bock, J., and Eckstein, L., "System Validation of Highly Automated Vehicles with a Database of Relevant Traffic Scenarios," *12th ITS European Congress*, Strasbourg, June 19-22, 2017.

26. Bustcon, J. and Randell, B. (Eds.), "Software Engineering Techniques: Report on a Conference Sponsored by the NATO Science Committee," April 1970.

27. Beizer, B., *Black-Box Testing: Techniques for Functional Testing of Software and Systems* (Wiley, 1995).

28. Zhou, N., "Volvo Admits Its Self-Driving Cars are Confused by Kangaroos," *The Guardian* (June 30, 2017), https://goo.gl/jgA7Ck.

29. Koopman, P., "Challenges in Autonomous Vehicle Validation," SCAV 17, April 2017.

30. Kane, Chowdhury, Datta, and Koopman, "A Case Study on Runtime Monitoring of an Autonomous Research Vehicle (ARV) System," RV, 2015.

31. Sargent, R., "Verifying and Validating Simulation Models," *2014 Winter Simulation Conference*, 118-131.

32. Law, A. and Kelton, W.D., *Simulation Modeling and Analysis*, 3rd ed. (McGraw-Hill, 2000).

33. Freedman, R., "Testability of Software Components," *IEEE Trans. Software Engineering* (June 1991): 553-564.

34. Dragan, A., Lee, K., and Srinivasa, S., "Legibility and Predictability of Robot Motion," *Human-Robot Interaction (HRI)* (2013): 301-308.

35. Bojarski, M. et al., "VisualBackProp: Efficient Visualization of CNNs," arXiv:1611.05418v3.

36. Bojarski, M. et al., "End to End Learning for Self-Driving Cars," arXiv:1604.07316v1.

37. Wang, Y., Lin, Z., Shen, X., Cohen, S. et al., "Skeleton Key: Image Captioning by Skeleton-Attribute Decomposition," arXiv preprint arXiv:1704.06972.

38. Redmon, J. and Farhadi, A., "YOLO9000: Better, Faster, Stronger," https://arxiv.org/pdf/1612.08242.pdf.

39. Morris, E., "The Certainty of Donald Rumsfeld (Part 2)," *NY Times*, March 26, 2014, https://goo.gl/Pv7SB7.

40. Wang, R., Guiochet, J., and Motet, G., "Confidence Assessment Framework for Safety Arguments," *SAFECOMP*, 2017, 55-68.

41. Casner, S., Hutchins, E., and Norman, D., "The Challenges of Partially Automated Driving," *Communications of the ACM* (May 2016): 70-77.

42. Leveson, "An Investigation of the Therac-25 Accidents," *IEEE Computer* (July 1993): 18-41.

43. Sullivan, M. and Chillarege, R., "Software Defects and Their Impact on System Availability a Study of Field Failures in Operating Systems," *FTCS-21*, 1991.

44. Kalra, N. and Groves, D., "The Enemy of Good: Estimating the Cost of Waiting for Nearly Perfect Automated Vehicles," Rand Corporation, RR-2150-RC, 2017.

45. Burton, S., "Making the Case for Safety of Machine Learning in Highly Automated Driving," *SAFECOMP*, September 2017, 5-16.

46. Kane, F. and Koopman, "Monitor Based Oracles for Cyber-Physical System Testing," DSN 2014.

Bayesian Test Design for Reliability Assessments of Safety-Relevant Environment Sensors Considering Dependent Failures

Mario Berk
Technical University of Munich/AUDI AG

Hans-Martin Kroll, Olaf Schubert, and Boris Buschardt
AUDI AG

Daniel Straub
Technical University of Munich

With increasing levels of driving automation, the perception provided by automotive environment sensors becomes highly safety relevant. A correct assessment of the sensors' perception reliability is therefore crucial for ensuring the safety of the automated driving functionalities. There are currently no standardized procedures or guidelines for demonstrating the perception reliability of the sensors. Engineers therefore face the challenge of setting up test procedures and plan test drive efforts. Null Hypothesis Significance Testing has been employed previously to answer this question. In this contribution, we present an alternative method based on Bayesian parameter inference, which is easy to implement and whose interpretation is more intuitive for engineers without a profound statistical education. We show how to account for different environmental conditions with an influence on sensor performance and for statistical dependence among perception errors. Additionally, we study the impact of error dependence among several sensors on the perception reliability of a redundant multi-sensor system. To this end, we simplify the sensor data fusion with a majority voting scheme,

CITATION: Berk, M., Kroll, H., Schubert, O., Buschardt, B. et al., "Bayesian Test Design for Reliability Assessments of Safety-Relevant Environment Sensors Considering Dependent Failures," SAE Technical Paper 2017-01-0050, 2017, doi:10.4271/2017-01-0050.

which implies that the multi-sensor system's perception fails whenever more than half of the individual sensors commit unacceptable errors. For a redundant multi-sensor system, in which error occurrence is weakly dependent, it can be shown that empirical reliability assessments are feasible. While the presented method does not encompass entirely the full complexity of the problem, it provides an initial systematic estimate of the necessary test drive effort and facilitates the use of sound statistical methods for test effort estimation.

Introduction

With the advent of advanced driver assistance systems (ADAS) and automated driving, machine vision provided by a set of environment perceiving sensors has become an integral part of modern cars [1, 2, 3, 4, 5, 6, 7, 8]. With this technological development arises the need to demonstrate the automated systems' safety and reliability before putting them in service. In this context it is important to assess the reliability of the environment sensors in perceiving the vehicles' surroundings because perception errors may have serious consequences.

The reliability of the automated system and of the sensors' perception depends strongly on the context and environment [9, 10, 11, 12]. Existing testing and validation frameworks, for instance ISO 26262, are not directly applicable to the safety validation of automated driving and machine vision [12]. This leads to the task of designing tests to validate the systems' safety. One important question for test design is: How much effort is necessary to empirically demonstrate the reliability of the sensor-based perception, i.e. how much real driving is necessary?

In the context of ADAS and machine vision, a statistical framework utilized to derive the empirical test effort is Null Hypothesis Significance Testing (NHST) [12, 13, 14, 15], which is based on the frequentist interpretation of probability. However, for many engineers with only basic training in statistics (and many scientists, see [16]), the correct interpretation of NHST is difficult and counterintuitive. There is a wide discussion in the scientific community about the misuse of NHST based on misinterpretation and overconfidence in "statistical significant" results [16, 17, 18, 19, 20]. In this contribution we therefore provide a Bayesian test design for empirical reliability assessments of environment perception. Bayesian methods for reliability assessments are widely used and common in many different industries [21, 22, 23, 24, 25, 26, 27].

One important advantage of the Bayesian method compared to the NHST is its flexibility, which allows easier application to non-standard problems such as the reliability of a correlated redundant multi-sensor system. With this method, the necessary test drive effort for a given safety target - e.g. on average less than one safety-relevant perception error in 10^8 hours - can be derived. We find that the Bayesian solution to this problem is easier to apply and interpret for engineers without advanced statistical training. The Bayesian test design is here applied to assess the perception reliability of environment sensors such as Lidar [28] and Radar [29]. Additional to the treatment of individual sensors, the perception reliability of a redundant multi-sensor system is in this study quantified with a majority-voting scheme that has been proposed in [15].

Statistical models utilized to estimate test efforts in the domain of ADAS rely on the assumption of independence of the critical events [12, 13, 14, 15]. In this contribution, perception error dependence is taken into account. The main contributions of this work are therefore:

1. The introduction of extensions to standard statistical models with which perception error dependence due to environmental effects can be accounted for at the individual sensor level.

2. The application of well-known and intuitive Bayesian methods to the problem of assessing the perception reliability of automotive environment sensors. The intention is to provide a detailed guide on perception reliability assessments for practicing engineers.

3. Statistical models for a majority-voting scheme that allow to consider error dependence between redundant sensors are presented in the context of estimating the perception reliability of a multi-sensor system.

The paper is organized as follows: First we review the idea of deriving the test drive effort for reliability demonstrations of ADAS and environment perception with NHST. Thereafter, we introduce a Bayesian approach as an alternative solution to the problem. It includes the dependence of subsequent perception errors and includes non-stationary perception error occurrences. Further, we address how to assess the perception reliability and safety of a redundant multi-sensor system, including error dependence among multiple sensors. A synthetic case study is conducted to demonstrate the proposed methods and to study the impact of error dependence on the reliability of environment perception provided by redundant sensors. Finally, a discussion of the results and the method is given and conclusions are provided.

Background: Reliability Assessment of Automotive Environment Perception

The aim of a reliability assessment is to demonstrate that the system or item under consideration - here at first the environment perception of an individual sensor - complies with a given target level of safety. In this specific case, the target level of safety can be expressed as an acceptable mean rate of safety-relevant perception error occurrences, denoted with λ_{SL}:

$$\lambda_{SL} = \frac{1}{\bar{t}} \tag{1}$$

where $\bar{t}$ is defined as the mean time between the occurrence of subsequent safety-relevant perception errors with the sensor under consideration. The definition of safety-relevant perception errors depends strongly on the ADAS or automated driving functionality of interest. Hence, the safety-relevance of an error has to be determined by the analyst. Generally, a safety-relevant perception error can be the non-detection of an object, the false-positive detection of an object, a large deviation of a physical measurement quantity from the ground truth (e.g. object position or object velocity) or a misclassification of the object (e.g. cyclist identified as pedestrian) [30]. With this definition of potential safety-relevant perception errors, an environment sensors' perception reliability refers to the probability of absence of safety-relevant perception errors, i.e. it is the compliment of the probability of safety-relevant perception error occurrence.

The target level of safety in Eq. (1) may be derived from norms. Alternatively, it was argued that reliability criteria may be defined by requiring the automated driving system to outperform human drivers in terms of the mean rate of accident occurrence [14].

Null Hypothesis Significance Testing for Sensor Reliability Assessment

NHST (see [31] for an introduction to statistics including NHST and [16, 17, 18, 19, 20] for the interpretation of NHST results) is a statistical tool utilized to test hypotheses and results in either making a decision in favor or against a tested hypothesis. To be able to make this decision, one sets up a so called null hypothesis H_0 which is opposed to the hypothesis of interest. The hypothesis of interest itself is termed alternative hypothesis H_1. The decision in favor of or against H_1 is based on whether the observed data of an experiment or test are unlikely to occur under the null hypothesis H_0.

When assessing the perception reliability of a sensor, a reasonable H_0 is that the sensor's mean safety-relevant perception error rate λ is larger than the desired target level of safety λ_{SL}:

$$H_0 : \lambda > \lambda_{SL} \tag{2}$$

The alternative hypothesis is that the sensor complies with the target level of safety:

$$H_1 : \lambda \le \lambda_{SL} \tag{3}$$

To be able to test H_1 with NHST, it is necessary to specify a random variable Z that summarizes the data (i.e. observations) of a test drive. Z is called the test statistic and is a function of the data. A particular observation of Z is denoted with z_{data}. An obvious choice for Z is the number of safety-relevant perception errors in a test drive. If under H_0, a value of smaller or equal to the observed z_{data} is unlikely to occur by chance (i.e. with a probability of less than α, where α denotes the significance level), it will be concluded that with a statistical confidence of $(1 - \alpha)$ the null hypothesis can be rejected. With this rejection one implicitly makes a decision in favor of H_1:

$$\text{Reject } H_0 \text{ if } \Pr\left(Z \le z_{data}|H_0\right) = p \le \alpha \tag{4}$$

p is the observed significance level of the data. The smaller the observed significance level p, the less likely it is for the observed z_{data} or smaller values to occur by chance, given H_0. It is important to understand that the decision to reject or not to reject H_0 based on $p \le \alpha$ is also conditional on the statistical assumptions underlying Eq. (4) (e.g. statistical model used, independence, data collection methods).

The necessary test drive effort that allows to make a decision in favor of $H_1 : \lambda \le \lambda_{SL}$ can then be derived by the following steps:

- Define a test statistic Z (here: the number of observed perception errors in the test drive)

- Derive the sampling distribution of Z

- Specify a significance level α (typically 0.05)

- Fix z_{data} at different values of the sample space (here: z_{data} is the number of observed errors, with possible values 0,1,2, ...)

- For each value of z_{data}, solve the underlying statistical model of Eq. (4) for the test effort (e.g. number of trials, time or kilometers) such that it holds $\Pr(Z \le z_{data}| H_0) = \alpha$.

Following these steps one obtains the minimum test drive effort associated with the acceptable number of perception errors, such that H_0 is rejected (for an analogous application see [14]).

When selecting a test design, i.e. a value of z_{data} and the corresponding test effort, the error of rejecting a true H_0 - termed type 1 error - is made with a probability α or less, conditional on H_0 and under the condition that the assumptions on the test statistics hold. The type 2 error occurs when not rejecting H_0, even though H_1 is the truth. In case of environment sensors, this error will occur if by chance more safety-relevant errors than acceptable are observed in the predefined test drive effort, even though $\lambda \leq \lambda_{SL}$. The probability β of the type 2 error can be quantified when assuming a specific error rate $\lambda \leq \lambda_{SL}$ to be the hypothetical truth. With a fixed test effort, simultaneously minimizing the type 1 and the type 2 errors is not possible. The lower the type 1 error should be, the larger the type 2 error becomes.

A major problem with NHST after rejecting H_0 is a common but flawed interpretation: Often it is concluded, that at least with $(1 - \alpha)$ probability H_1 is true, or with $(1 - \alpha)$ probability H_0 is not true [16]. This interpretation is wrong [16, 17, 18, 19, 20]. A specific successful hypothesis test that rejects $H_0 : \lambda > \lambda_{SL}$ does not demonstrate $H_1 : \lambda \leq \lambda_{SL}$ with a certain probability nor does the p-value specify the probability of the data occurring by chance [20]. The hypothesis test only allows the statement that, given the data fulfills Eq. (4), it is not a bad decision to reject H_0, because if H_0 was true, then the observed data would be unlikely. A single hypothesis test makes no statement about the probability of neither H_0 nor H_1 being the truth; in the context of NHST these probabilities are either zero or one [16].

The next section discusses the implications of using NHST for reliability assessments and puts the presented misconception into perspective.

PERFORMANCE EVALUATION OF NHST

From a societal risk perspective an important question one might ask is: What is the probability of releasing a system with $\lambda > \lambda_{SL}$ (i.e. one that does not comply with the target level of safety), when using the NHST method? The answer to this question depends on how many systems tested with NHST fulfill $\lambda \leq \lambda_{SL}$ in the first place, i.e. on the prior probability $\Pr(H_1)$, and can be estimated as [32]:

$$\Pr\left(H_0 | rejection\ of\ H_0\right) = \frac{\left[1 - \Pr\left(H_1\right)\right] \cdot \alpha}{\Pr\left(H_1\right) \cdot \left(1 - \beta\right) + \left[1 - \Pr\left(H_1\right)\right] \cdot \alpha} \qquad (5)$$

where $\Pr(H_0 | rejection\ of\ H_0)$ denotes the ratio of released systems not complying with the target level of safety λ_{SL} to total systems released in the long run. Eq. (5) with its dichotomization of the error rate λ into H_0 and H_1 is in fact a simplification of a continuous probabilistic problem. Therefore, Eq. (5) holds approximately if systems not complying with the target level of safety (i.e. the systems for which H_0 holds) have a λ in the unsafe region $\lambda > \lambda_{SL}$ close to λ_{SL}, and if all systems that fulfill the target level of safety have the same type 2 error probability β. For the case of $\beta = 0.5$, the ratio of released systems with $\lambda > \lambda_{SL}$ to the total number of systems released, after Eq. (5), is illustrated in Figure 1 as a function of the prior probability $\Pr(H_1)$.

First, as Figure 1 shows, if no system complies with the target safety level λ_{SL} in the first place, i.e. $\Pr(H_1) = 0$, then all systems released with NHST fail to comply with the target level of safety. This is a trivial result, but is pointed out here considering the possible misinterpretations of the p-value. Second, if for instance 20% of the systems tested comply with the target level of safety, i.e. $\Pr(H_1) = 0.2$, then roughly 30% of all systems released are erroneously considered to comply with the predefined safety requirements. This is far from the significance level $\alpha = 0.05$, which demonstrates how the true error of NHST is easily underestimated when α is misinterpreted to be a probabilistic statement about the tested hypotheses.

In reality, $\Pr(H_1)$ is unknown and the true percentage of released systems not complying with the target level of safety cannot be known with certainty. Nevertheless,

Percentage
Pr(H_0|*rejection of H_0*) of released
systems with NHST that do not comply
with the target level of safety, in function
of the prior probability Pr(H_1) for a
system complying with the target level
of safety, assuming a type 2 error of
$\beta = 0.5$ and a significance level $\alpha = 0.05$.

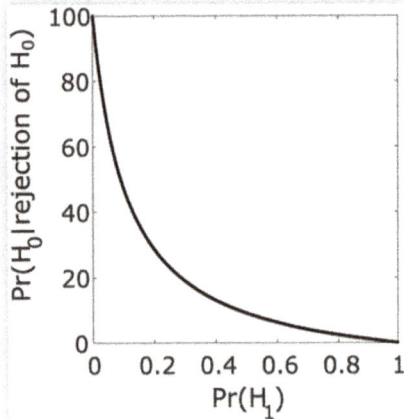

Figure 1 demonstrates how one should not be overconfident in NHST results that show the system under consideration is reliable with "statistical significance".

ALTERNATIVES TO NHST FOR RELIABILITY ASSESSMENTS

Testing engineers and all involved stakeholders of a reliability analysis might benefit from conceptually easier and more transparent methods than NHST when trying to demonstrate the reliability of new systems such as ADAS or environment sensors. The American Statistical Association (ASA) recently issued a warning about p-values of NHST due to their misuse and misinterpretation [20]: "Scientific conclusions and business or policy decisions should not be based only on whether a p-value passes a specific threshold." Confidence intervals are often put forward as an alternative to NHST. They are related to concepts of NHST and are equally likely to be misinterpreted [16, 33]. Therefore we refrain from further discussing this option.

We find that the Bayesian view on probability is in most contexts better suited for empirical sensor perception reliability evaluation. In contrast to frequentist approaches such as NHST, the Bayesian interpretation of probability treats observed data as fixed and the probabilistic parameters that produced the data as random. The Bayesian approach is often conceptually easier and more directly answers the question usually asked by the analyst: Which probabilistic conclusions about an uncertain parameter of interest can be drawn from a particular set of data or observations [17, 34]? For some problems, a frequentist analysis may yield the same numerical results as a Bayesian analysis [35] but, strictly speaking, does not allow the same intuitive interpretation. The interested reader is referred to [34, 35, 36] for further information on the frequentist and Bayesian point of view of probability.

Bayesian Methodology for Empirical Perception Reliability Assessments of Environment Sensors

In this section, we propose a Bayesian alternative to NHST to derive the necessary test effort for reliability assessments of environment perception in the field of ADAS. First, a statistical model is presented that accounts for dependent perception errors and for a non-stationary probability of error occurrence. Following the definition of the statistical model, a Bayesian solution to estimate the test drive effort and to perform an empirical sensor perception reliability assessment is derived. Moreover, statistical models are presented that allow to assess the reliability of a multi-sensor system.

Statistical Model

Environment sensors such as Lidars [28] and Radars [29] repeatedly probe their environment in measurement cycles and aggregate the collected information in a time-discretized digital environment model containing relevant information about the driving environment and the traffic participants [30]. In this section, the focus is on individual sensors, i.e. the environment representation is not yet based on sensor data fusion [37] but on the data of an individual sensor. Given the temporal discretization of the environmental model, a natural way of describing error occurrence of an individual sensor is to introduce a binary random variable W_i for each measurement cycle i: Either the measurement cycle i is free from safety-relevant perception errors ($w_i = 0$) or at least one safety-relevant perception error occurs ($w_i = 1$). All different perception error types defined previously in the section *Background: Reliability assessment of automotive environment perception* are here considered jointly with the random variable W_i.

With this interpretation, the occurrence of safety-relevant perception errors ($w_i = 1$) in a single measurement cycle i is represented by a Bernoulli trial:

$$pw_i(w_i) = \begin{cases} p & for & w_i = 1 \\ 1-p & for & w_i = 0 \end{cases} \tag{6}$$

where p is the probability of error occurrence. The probability of the number of safety-relevant perception errors Y in n measurement cycles can then be modeled with the Binomial distribution:

$$p_Y(y) = \binom{n}{y} \cdot p^y \cdot (1-p)^{n-y} = \cdots$$
$$= \frac{n!}{y! \cdot (n-y)!} \cdot p^y \cdot (1-p)^{n-y} \tag{7}$$

Whenever the probability of error occurrence p is small ($p \to 0$) and the number of measurement cycles is large ($n \to \infty$), both of which holds for environment sensors, in the limit as $n \to \infty$, the Binomial distribution leads to the Poisson distribution:

$$p_{X_t}(x) = \frac{(\lambda \cdot t)^x}{x!} \cdot exp(-\lambda \cdot t) \tag{8}$$

where $x \in [0, 1, 2, \ldots]$ is the number of safety-relevant perception errors in the time interval t and λ is the mean rate of safety-relevant perception error occurrence.

In order for Eqs. (6), (7), 8) to hold, two important requirements have to be met: First, error occurrences in subsequent measurement cycles have to be independent of each other, and second, the probability p and thus the error rate λ have to be constant. Both requirements are not met for environment sensors. The performance of environment sensors such as Lidars or Radars depends on the given context and external factors, including adverse weather conditions, dirt, dust and target properties [9, 10, 11, 12]. As a consequence, p and λ are not constant over time. Also, if an error occurs in a given measurement cycle, it will be more likely for the subsequent measurement cycle to exhibit an error due to common influencing factors. Therefore, error occurrence is not independent of each other. Thus, the distributions provided by Eqs. (6), (7), 8) cannot be utilized without violating the underlying mathematical assumptions.

MATHEMATICAL REPRESENTATION OF DEPENDENT ERRORS

The two violations discussed in the previous section are seen to be caused by physical effects that act on different time scales. The perception error dependence is caused by physical effects that are common to multiple measurement cycles in a row. Examples are the presence of objects with low reflectivity, strong rain gusts or a low sun that blinds optical sensors. Due to the highly dynamic nature of driving vehicles, the effects causing the dependence are often only present for a short duration, in the scale of a few seconds. Environment conditions with influence on sensor performance which act on a scale in the order of minutes to hours, such as the weather in general, are not seen as the primary cause of dependent errors but rather influence the overall probability of error occurrence in a specific time interval. These effects consequently lead to a non-stationary error rate and are treated in the next section.

The error dependence leads to a higher probability of error occurrence in subsequent measurement cycles, once an error has occurred. If one wanted to estimate the probability of an perception error occurring for two measurement cycles in a row, with the model given in Eq. (7), neglecting the dependence could lead to severe underestimation. Aside of violating the requirements for Eqs. (6), (7), 8), error dependence is an important factor to consider when assessing the perception reliability of environment sensors because the safety-relevance of errors is partly determined by whether errors persist over multiple cycles (e.g. a false-positive object). A perception error occurring in only one cycle does typically not lead to an insecure or inappropriate behavior of a desired functionality. Sensors are able to use a multi-cycle validation, restricting the impact of errors occurring only for a very short duration [15]. This means that ADAS such as collision protection systems or adaptive cruise control only react when information is consistent over multiple measurement cycles [15, 29, 38].

To account for dependent errors caused by physical effects such as outlined above, the reference of the mean rate of error occurrence λ has to be adapted. In Eq. (8), λ is the rate of safety-relevant errors referring to individual measurement cycles. To consider dependent errors, the error rate λ of Eq. (8) is associated with the interpretation given in Figure 2. F_1, F_2 and F_3 (and so on) are the events that subsequent measurement cycles contain at least one error, at least two errors and at least three errors in a row. Accordingly, λ_1, λ_2 and λ_3 refer to the rate of occurrences of at least one error, at least two errors and at least three errors in a row. Additionally, F_0 denotes the event that a single measurement cycle is free from perception errors. Due to the safety relevance of perception errors that persist over multiple cycles, the analyst ultimately is not interested in λ_1 but rather in λ_2, λ_3 or the rates associated with a larger number of subsequent events.

It should be clear, because $\lambda_j \leq \cdots \leq \lambda_2 \leq \lambda_1$, it is easier to learn λ_1 than e.g. λ_2 as more data will be available for λ_1 than λ_2. With the interpretation given in Figure 2, for a fixed number of measurement cycles n, it holds:

$$n = n_{F_0} + n_{F_1} + n_{F_2} + \cdots + n_{F_\infty}$$ (9)

Where n_{F_0} are the number of F_0 events, n_{F_1} are the number of F_1 events and so on. When n becomes large ($n \to \infty$), n_{F_1} is related to n_{F_0}:

$$n_{F_1} = n_{F_0} \cdot \Pr\left(F_1|F_0\right)$$ (10)

FIGURE 2   Each box represents a sensor's measurement cycle. A grey colored box indicates that a perception error has occurred in the given cycle. F_1, F_2, F_3 are the events that at least one, at least two and at least three cycles in a row contain an error. λ_1, λ_2, λ_3 are the rates of error occurrences referring to the events F_1, F_2, F_3. F_0 denotes that no error has occurred in a given cycle.

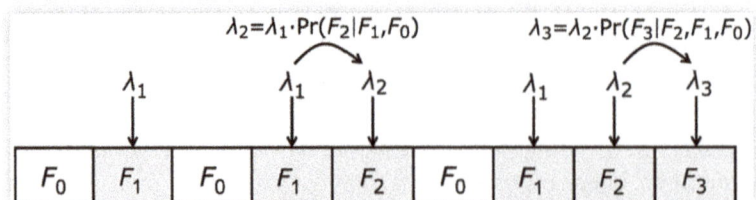

where $\Pr(F_1|F_0)$ is the probability that the first cycle of a potential row of errors contains an error, given no error has occurred in the previous cycle (see Figure 2). Similarly, n_{F_j} can be obtained:

$$n_{F_j} = n_{F_0} \cdot \prod_{i=1}^{i=j} \Pr\left(F_i|F_{i-1,\dots,}F_0\right) \tag{11}$$

$\Pr(F_i|F_{i-1}, \dots, F_0)$ is the probability of i errors in a row, given $i-1$ errors in a row have occurred previously. Thus the dependence of error occurrence is quantified with $\Pr(F_i|F_{i-1}, \dots, F_0)$ for all $i = 1, 2, 3, \dots$. Inserting Eqs. (10) (11) into Eq. (9) and solving for n_{F_0}/n leads in the limit of $n \to \infty$ to the unconditional probability $\Pr(F_0)$ of a randomly selected cycle to be free from perception errors:

$$\Pr\left(F_0\right) = \lim_{n\to\infty}\frac{n_{F_0}}{n} =$$

$$= \frac{1}{1 + \Pr\left(F_1|F_0\right) + \Pr\left(F_1|F_0\right)\cdot\Pr\left(F_2|F_1,F_0\right) + \dots + \prod_{j=1}^{j=\infty}\Pr\left(F_j|F_{j-1},\dots,F_0\right)} \tag{12}$$

Based on $\Pr(F_0)$, λ_1 is obtained as:

$$\lambda_1 = \frac{\Pr\left(F_0\right)\cdot\Pr\left(F_1|F_0\right)}{t_{cycle}} \tag{13}$$

With t_{cycle} the measurement cycle time. Generally, it holds:

$$\lambda_j = \lambda_{j-1}\cdot\Pr\left(F_j|F_{j-1},\dots,F_0\right) \tag{14}$$

It follows from Eqs. (12) (14) that λ_j fully describes the dependence structure which is quantified by $\Pr(F_0)$, $\Pr(F_1|F_0)$,.., $\Pr(F_j|F_{j-1}, \dots, F_0)$. Therefore, if the interest is in a sequence of at least j errors in a row, the dependence is fully accounted for. Furthermore, as long as the time intervals are large ($t \to \infty$) and the error events are rare, the number of F_j events in two non-overlapping time intervals can for a given λ_j be regarded as approximately independent of each other. Both these requirements can be assumed to hold for environment sensors. Under these conditions Eq. (8) can be used. Another way of interpreting λ_j and λ_{j-1} in Eq. (14) is that they are related by a Poisson process with random selections [39]. In the remainder of the contribution the index j of the λ_j of interest will be dropped for ease of notation.

Considering a Non-Stationary Error Rate

In this section, the non-stationary rate of error occurrence on a larger time scale is addressed. Environmental conditions and effects with influence on sensor performance that lead to a non-stationary rate of error occurrence are for instance adverse weather, dust, dirt, temperature and many more [9, 10, 11, 12].

To account for the non-stationary rate of error occurrence, $\lambda \cdot t$ is in Eq. (8) is replaced by $\mu(t)$:

$$p_{X_t}\left(x\right) = \frac{\mu\left(t\right)^x}{x!}\cdot exp\left(-\mu\left(t\right)\right) \tag{15}$$

$\mu(t)$ is the mean number of safety-relevant errors in the time interval t. For Lidar sensors, weather influences are among the most relevant environmental effects [10, 11].

We therefore use the example of weather conditions to present the calculation of $\mu(t)$. Let the weather be characterized by the four conditions sunny, rainy, snowy and cloudy weather. The sensor performance might differ under different conditions. When the time interval t is large, the mean number of error occurrence $\mu(t)$ can be approximated as:

$$\mu(t) = (p_{sun} \cdot \lambda_{sun} + p_{rain} \cdot \lambda_{rain} + p_{snow} \cdot \lambda_{snow} + \cdots \\ + p_{cloudy} \cdot \lambda_{cloudy}) \cdot t \tag{16}$$

where p_{sun}; p_{rain}; p_{snow} and p_{cloudy} are the probabilities of sunny, rainy, snowy and cloudy weather. λ_{sun}, λ_{rain}, λ_{snow}, λ_{cloudy} are the mean rates of error occurrence during rainy, snowy and cloudy weather. The average error rate $\bar{\lambda}$ can then be calculated as:

$$\bar{\lambda} = p_{sun} \cdot \lambda_{sun} + p_{rain} \cdot \lambda_{rain} + p_{snow} \cdot \lambda_{snow} + \cdots \\ + p_{cloudy} \cdot \lambda_{cloudy} \tag{17}$$

It has to hold $p_{sun} + p_{rain} + p_{snow} + p_{cloudy} = 1$. Additional environmental effects can be considered by decomposing each error rate λ_{sun}, λ_{rain}, λ_{snow}, λ_{cloudy} with respect to the environmental effect that should be added to the model, in analogy to Eq. (17).

To correctly estimate a representative $\bar{\lambda}$, the test drive of total duration t has according to Eqs. (16) (17) be conducted in accordance with $t_{sun} = p_{sun} \cdot t$; $t_{rain} = p_{rain} \cdot t$; $t_{snow} = p_{snow} \cdot t$; $t_{cloudy} = p_{cloudy} \cdot t$. Note that under a varying rate of error occurrence, the error occurrences no longer follows a Poisson process. Nevertheless, the probability of the number of error occurrences can be described by Eq. (8), in which $\lambda \cdot t$ is replaced by $\mu(t)$ (or equivalently by replacing λ with $\bar{\lambda}$).

A problem is however that the probability of for instance p_{rain} varies geographically. For now it is assumed one is able to learn the probabilities p_{sun}; p_{rain}; p_{snow} and p_{cloudy} for one geographical region. Then, for this particular region, the non-stationary rate of error occurrence is accounted for when the test drive is conducted in accordance with p_{sun}; p_{rain}; p_{snow} and p_{cloudy}. A more detailed examination of how to learn λ_{sun}, λ_{rain}, λ_{snow}, λ_{cloudy} individually and independent of the geographical region in an efficient way is beyond the scope of this contribution.

Bayesian Reliability Assessment and Test Effort Estimation

This section describes a Bayesian method for deriving the necessary test drive effort to demonstrate $\lambda < \lambda_{SL}$ before the data is collected and for assessing the reliability of the sensor after the test drive, once the data is available. The general problem is that of inferring an unknown mean rate of safety-relevant perception error occurrence λ from a limited amount of data, where the data consists of the number of safety-relevant perception errors x that have been observed in a specific time interval t. We use Bayesian statistics (see [40] for an introduction) to solve this problem. For a detailed treatment of Bayesian reliability analyses we refer to textbooks [21, 22].

Bayes' theorem is applied to draw probabilistic conclusions on λ:

$$f(\lambda \mid x, t) \propto f(\lambda) \cdot p_{Xt}(x \mid \lambda, t) \tag{18}$$

$f(\lambda \mid x, t)$ is the posterior probability distribution of the safety-relevant perception error rate λ for a given observed number of safety-relevant errors x in the time interval t, $f(\lambda)$ is the prior probability distribution of the error rate λ and $p_{Xt}(x \mid \lambda, t)$ is the likelihood of λ given the observation of x in t. The likelihood $p_{Xt}(x \mid \lambda, t)$ is defined by the

Poisson distribution of Eq. (8). The symbol $\propto$ in Eq. (18) expresses that the posterior distribution is proportional to the prior and likelihood up to a constant.

A convenient choice for the prior distribution in case of a Poisson likelihood is the Gamma distribution. The Gamma distribution is the conjugate distribution to the Poisson likelihood, which signifies that both $f(\lambda)$ and $f(\lambda|x,t)$ in Eq. (14) have the Gamma distribution [40]. The Gamma probability density function (PDF) is:

$$f(\lambda) = \frac{b^a}{\Gamma(a)} \cdot \lambda^{a-1} \cdot \exp(-b \cdot \lambda) \tag{19}$$

Where a and b are the parameters of the gamma distribution and $\Gamma(a) = \int_0^\infty u^{a-1} \cdot \exp(-u) du$ is the gamma function. The corresponding Gamma cumulative distribution function (CDF) $F(\lambda|x,t)$ is:

$$F(\lambda|x,t) = \frac{\gamma(a,b \cdot \lambda)}{\Gamma(a)} \tag{20}$$

Where $\gamma(a,b \cdot \lambda) = \int_0^{b \cdot \lambda} u^{a-1} \cdot \exp(-u) du$ is the incomplete gamma function. The prior distribution is described by $f(\lambda)$ with parameters a' and b'. The parameters of the posterior $f(\lambda|x,t)$ are denoted a'' and b'' and are obtained as:

$$a'' = a' + x \tag{21}$$

$$b'' = b' + t \tag{22}$$

Inserting λ_{SL} together with a'' and b'' into Eq. (20) provides the answer to the key question in this probabilistic reliability assessment: What is the probability $\Pr(\lambda < \lambda_{SL}|x,t)$ that the system under consideration complies with the target level of safety λ_{SL}?

$$\Pr(\lambda < \lambda_{SL}|x,t) = F(\lambda_{SL}|x,t) \tag{23}$$

Moreover, the best point estimate of the unknown error rate λ is the posterior mean $\hat{\lambda}$:

$$\hat{\lambda} = \frac{a''}{b''} = \frac{a' + x}{b' + t} \tag{24}$$

As the analysis deals with a safety-relevant issue, often a more conservative estimate for λ than the posterior mean $\hat{\lambda}$ is desired. Therefore the analyst may chose for instance the 95% quantile, or alternatively the 99% quantile, of the posterior λ as a conservative point estimate.

To perform the analysis, prior parameters have to be selected. A commonly accepted formal rule to construct an (objective) prior distribution when no prior information is available has been defined by Jeffreys [40, 41, 42]. The property that makes Jeffreys' prior non-informative is its invariance to re-parameterizations [40]. Here, Jeffreys' prior yields $a' = 0.5$ and $b' \to 0$. Eq. (24) supports the interpretation of the prior parameters as a' prior error observations in a prior test time interval b' (see [21] page 89). However, by comparing Eq. (19) with Eq. (8), Gelman et al. [40] page 52 argue that the prior parameters may be interpreted as a' - 1 prior observations in a prior time interval b'. Following this interpretation, a weakly informative prior in case no prior information is available could also be selected as $a' = 1$ and $b' \to 0$. One is able to show that with $a' = 1$ and $b' \to 0$ the same numerical results for the necessary test effort t are obtained as with NHST. If substantial information prior to the analysis is available, then this information can easily be incorporated into a' and b' following the interpretation given.

The test drive effort before collecting the data can be derived by the following steps:

- Select the probability with which the target level of safety should be complied with (e.g. $\Pr(\lambda < \lambda_{SL}|x,t) = 0.95$)

- Insert the desired target level of safety λ_{SL} into Eq. (20)

- Fix the acceptable number of errors x of the test drive at different values (i.e. $x = 0$, 1, 2, ...)

- For each value of x, solve Eq. (20) for the test effort t

The result are the acceptable number of errors x for a given test drive effort t, which all allow to conclude with at least 95% probability that the sensor complies with λ_{SL}.

Assessing the Reliability of a Multi-Sensor System

To enhance the safety of the environment perception, the system architecture may include redundant sensors obtaining the same types of information in overlapping field of views. It is pointed out that complementary sensor principles used to obtain different types of information (e.g. camera for object classification and radar for object localization) do not comprise a redundant but a complementary system [37]. These are not considered here.

To combine the information and data of multiple sensors, sensor data fusion is applied [43]. Modern sensor data fusion algorithms mostly are based on Bayes filters such as the well-known Kalman filter [43]. Because it is not straightforward to evaluate the performance of a multi-sensor system with complicated fusion algorithms, we simplify the problem with a so-called majority voting system [12, 15]. The assumption is that a system's perception using the information of redundant sensors fails, when more than half of the individual sensors commit safety-relevant perception errors. An approach utilizing a majority voting scheme to describe the multi-sensor based perception reliability has already been reported in [15].

In reliability analysis, a majority voting system can be represented as a k-out-of-N system [22], meaning that at least k-out-of-N sensors have to commit safety-relevant perception errors for the system to provide erroneous information. To calculate the multi-sensor system's rate of perception error occurrence, let the occurrence of safety-relevant errors for each sensor $s = 1, \ldots, N$ (N is the total number of redundant sensors) for a given measurement cycle be a binary random variable U_s with $u_s = 1$ meaning sensor s commits at least one error and $u_s = 0$ meaning sensor s commits no error. The probability p of committing an error is assumed to be equal for all sensors and is related to the error rate λ of the individual sensors as well as the measurement cycle time t_{cycle}:

$$p = 1 - \exp\left(-\lambda \cdot t_{cycle}\right) \approx \lambda \cdot t_{cycle} \tag{25}$$

The approximation $p = \lambda \cdot t_{cycle}$ holds for $\lambda \ll 1$ h^{-1}. The multi-sensor machine vision, based on majority voting, commits perception errors when $\sum_{s=1}^{N} U_s \geq \left\lfloor \frac{N}{2} + 1 \right\rfloor$, with $\left\lfloor \frac{N}{2} + 1 \right\rfloor$ being the notation for rounding $\frac{N}{2} + 1$ down. Under the assumption that the individual sensors' perception error probabilities p are independent of each other, the probability p_f of the multi-sensor based machine vision to provide erroneous information can be calculated with the binomial CDF:

$$p_f = \sum_{k=\left\lfloor \frac{N}{2}+1 \right\rfloor}^{N} \binom{N}{k} \cdot p^k \cdot \left(1-p\right)^{N-k} \tag{26}$$

Adverse physical conditions such as the presence of objects with low reflectivity or strong rain gusts might lead to dependence between potential safety-relevant perception errors of redundant sensors (equivalent to the discussion in the previous section). Therefore the assumption of independence in Eq. (26) might not be justified. To take dependent multi-sensor errors into account, we define the correlation coefficient ρ of perception error occurrence U_s and U_q between any pairs of sensors $s, q \in \{1, \dots, N\}$ [44]:

$$\rho = \frac{E\left[U_s \cdot U_q\right] - E\left[U_s\right] \cdot E\left[U_q\right]}{\sqrt{E\left[U_s\right]\left(1 - E\left[U_s\right]\right) \cdot E\left[U_q\right] \cdot \left(1 - E\left[U_q\right]\right)}} \tag{27}$$

where E[ ]denotes the expectation operator. An important aspect is the interpretation of the correlation coefficient given by Eq. (27), which in this form is not very intuitive. When identical sensors are utilized, it holds $E[U_s] = E[U_q] = p$ and further:

$$\begin{aligned} E\left[U_s \cdot U_q\right] &= \Pr\left(U_s = 1 \cap U_q = 1\right) = \cdots \\ &= \Pr\left(U_s = 1 | U_q = 1\right) \cdot \Pr\left(U_q = 1\right) = \cdots \\ &= \Pr\left(U_s = 1 | U_q = 1\right) \cdot p \end{aligned} \tag{28}$$

Inserting into Eq. (27) leads to:

$$\rho = \frac{\Pr\left(U_s = 1 | U_q = 1\right) \cdot p - p^2}{p - p^2} \tag{29}$$

Eq. (29) is now more easily interpreted: $\Pr(U_s = 1| U_q = 1)$ is the conditional probability that sensor s commits a perception error given that sensor q has committed an error. p is the individual sensors' probability of perception error occurrence. On the one hand, with independence, i.e. $\Pr(U_s = 1| U_q = 1) = \Pr(U_s = 1) = p$, the correlation coefficient becomes $\rho = 0$. On the other hand, if it is certain that sensor s commits an error when an error occurs in sensor q, i.e. $\Pr(U_s = 1| U_q = 1) = 1$, the correlation coefficient becomes $\rho = 1$. This is equivalent to full dependence. Finally, when p is small, it holds $p^2 \ll p$ and Eq. (29) can be simplified to:

$$\rho \approx \Pr\left(U_s = 1 | U_q = 1\right) \tag{30}$$

That is, the correlation coefficient ρ is approximately equal to the conditional probability of error occurrence in sensor s given an error has occurred in sensor q.

To account for perception error dependence between redundant sensors we consider the beta-binomial distribution [44, 45, 46, 47] and a model for correlated binary data proposed by Gupta and Tao [48]. The latter model is introduced because the beta-binomial distribution can due to numerical reasons not be utilized with small values of the correlation coefficient ρ. Conversely, the Gupta and Tao model is not applicable to large values of correlation ρ (for further information see [49, 50]). The exact values of ρ up to which both models can be used depend on the probability p. In the subsequent case study we utilize the beta binomial model for $\rho \geq 0.01$ and the Gupta and Tao model for $\rho < 0.01$.

With the beta-binomial distribution, the probability $\Pr\left(\sum_{s=1}^{N} U_s = k\right)$ of exactly k-out-of-N sensors committing perception errors is [46]:

$$\Pr\left(\sum_{s=1}^{N} U_s = k\right) = \binom{N}{k} \frac{\Gamma\left(\theta_1 + \theta_2\right) \cdot \Gamma\left(\theta_1 + k\right) \cdot \Gamma\left(\theta_2 + N - k\right)}{\Gamma\left(\theta_1\right) \cdot \Gamma\left(\theta_2\right) \cdot \Gamma\left(\theta_1 + \theta_2 + N\right)} \tag{31}$$

with $\Gamma(\ )$ the gamma function and the parameters θ_1, θ_2 related to the individual sensors' probability of perception error occurrence p and the correlation coefficient ρ (derived from [47]):

$$\theta_1 = \frac{p \cdot (1-\rho)}{\rho}, \quad \theta_2 = \frac{(1-p) \cdot (1-\rho)}{\rho} \tag{32}$$

Both p and ρ are assumed to be the equal for all sensors $s = 1, \ldots, N$ and p is given by Eq. (25) when the error rate λ of an individual sensor is known. The probability of the majority vote based multi-sensor machine vision to commit a perception error p_f, accounting for dependent sensor errors, can then be calculated as:

$$pf = \sum_{k=\left\lfloor \frac{N}{2}+1 \right\rfloor}^{N} \Pr\left(\sum_{s=1}^{N} U_s = k \right) \tag{33}$$

Eq. (25), (31) and (33) allow to study the system's perception reliability including error dependence among redundant sensors according to the beta-binomial model. The solution utilizing the Gupta and Tao [48] model is given in the appendix.

In theory, when the correlation coefficient goes to $\rho \to 0$, both the beta-binomial distribution as well as the Gupta and Tao model converge to the (independent) binomial distribution [46, 48]. Thus with $\rho \to 0$ the multi-sensor probability of perception error occurrence converges to Eq. (26), independently of which of the two models is utilized.

Case study: Empirically Demonstrating the Perception Reliability of Environment Sensors

A synthetic case study is performed. Suppose one is interested in demonstrating that a sensor's environment perception fulfills the target level of safety $\lambda_{SL} = 10^{-7}\,\mathrm{h}^{-1}$. For this target, first the necessary test drive effort t to demonstrate $\lambda < \lambda_{SL}$ is derived. With the derived test effort it is demonstrated how the test drive has to be conducted to account for different weather influences. It is assumed that the weather in a particular region can be described with the probabilities $p_{sun} = 0.65$; $p_{rain} = 0.15$; $p_{snow} = 0.05$ and $p_{cloudy} = 0.15$.

Thereafter, we draw probabilistic conclusions about the safety-relevant perception error rate λ based on hypothetical results of a test drive with the aim of illustrating the uncertainty in estimating λ.

In the last section of the case study, we study the influence of dependence on perception error occurrence in a multi-sensor based machine vision system. To this end, we assume that a specific task of the machine vision (e.g. object detection in the front field of view) is based on three identical Lidar or Radar sensors, respectively. The individual error rate λ of the three sensors is thus identical. The measurement cycle time is assumed to be $t_{cycle} = 0.05$ s.

For all calculations in this case study, Jeffreys' prior parameters $a' = 0.5$ and $b' = 0$ are selected in Eq. (18). Jeffreys' prior reflects ignorance on the error rate λ before conducting the test and is commonly considered to be non-informative [40]. This choice of prior is conservative as it only assigns a prior probability of $\Pr(\lambda < \lambda_{SL} = 10^{-7}\,\mathrm{h}^{-1}) = 1.13 \cdot 10^{-11}$ that the target level of safety is met.

Estimating the Necessary Test Drive Effort

The empirical test drive effort can be estimated with Eq. (23). Figure 3a) shows the probability $\Pr(\lambda < \lambda_{SL} | x, t)$ that the sensor under consideration complies with the target level of safety $\lambda_{SL} = 10^{-7}$ h^{-1} for the cases of $x = 0$, $x = 1$ and $x = 2$ perception errors during a test drive, in function of the test drive effort t. To derive a conservative estimate of the test drive effort t required for a demonstration of $\lambda < \lambda_{SL}$, one may select the t for which it holds $\Pr(\lambda < \lambda_{SL} | x, t) = 0.95$ as indicated for the case $x = 0$ in Figure 3a) with the grey arrow.

A summary of different combinations of test drive efforts t and the corresponding acceptable number of errors x that all allow to conclude $\Pr(\lambda < \lambda_{SL} | x, t) = 0.95$ are given in Figure 3b). The smallest possible test effort to demonstrate $\lambda < \lambda_{SL}$ requires the observation of no errors and is $t = 1.92 \cdot 10^7$ h. Even though all combinations in Figure 3b) show compliance with the target level of safety with 95% credibility, it is necessary to select the test drive effort a-priori, and not do adjust it based on the observed number of errors. If a stopping criteria is selected on the go, the estimate of $\Pr(\lambda < \lambda_{SL} | x, t)$ is biased.

Suppose the test effort is selected as $t = 1.92 \cdot 10^7$ h with no acceptable errors x during the test. To consider that error rates might differ substantially in function of environmental conditions with influence on sensor performance, the test effort has to be distributed according to the probabilities of the different environment conditions. Here only weather conditions are considered as relevant influencing factors. The resulting test profile is summarized in Table 1. Application of this test profile takes non-stationary error rates according to Eq. (17) into account.

TABLE 1 Resulting test profile to account for different weather conditions

Weather condition	Test time t_i for weather condition i
Sunny	$t_{sun} = 1.248 \cdot 10^7$ h
Rain	$t_{rain} = 0.288 \cdot 10^7$ h
Snow	$t_{snow} = 0.096 \cdot 10^7$ h
Cloudy	$t_{cloudy} = 0.288 \cdot 10^7$ h

FIGURE 3 a) Probability $\Pr(\lambda < \lambda_{SL} | x, t)$ of compliance with the target level of safety $\lambda_{SL} = 10^{-7}$ h^{-1} in function of the test drive effort t, for the cases of $x = 0$, $x = 1$ and $x = 2$ errors during the test. The grey arrow indicates the test drive effort for the case of $x = 0$ errors and $\Pr(\lambda < \lambda_{SL} | x, t) = 0.95$. b) Number of acceptable errors x in a test drive as a function of the test drive effort t such that $\Pr(\lambda < \lambda_{SL} | x, t) = 0.95$.

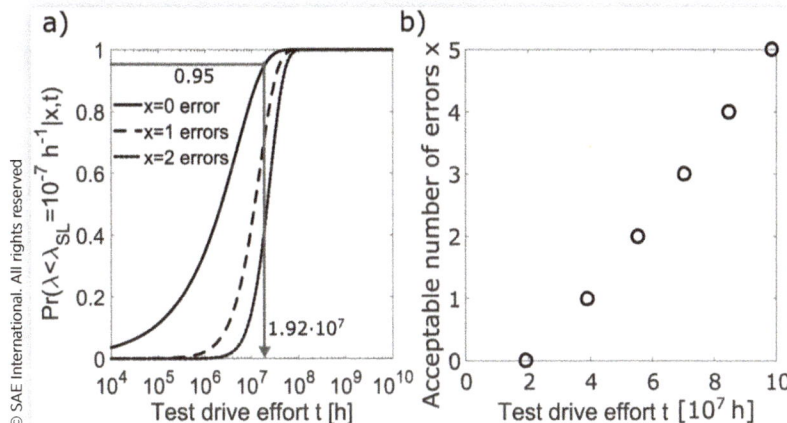

Evaluating Hypothetical Test Results

In this section it is assumed that a test drive with $t = 1.92 \cdot 10^7$ h has been conducted. Two hypothetical results of this test drive are evaluated: (a) $x = 0$ and (b) $x = 1$ errors have been observed in t. With these test results, the posterior parameters of the gamma distribution are calculated according to Eq. (21) and Eq. (22) : (a) $a'' = 0.5$, $b'' = 1.92 \cdot 10^7$ and (b) $a'' = 1.5$, $b'' = 1.92 \cdot 10^7$. The resulting posterior PDFs and CDFs of the error rate λ with these parameters are illustrated in Figure 4.

As the PDFs in Figure 4a) and Figure 4c) show, the Bayesian approach captures the full uncertainty in the error rate λ by assigning to each value of the error rate a probability density. The CDF Figure 4b) shows for $\lambda = 0$ that the unknown error rate is with 95% probability smaller than $\lambda_{SL} = 10^{-7}$ h^{-1}. This is in accordance with the test design derived in the previous section. With the observation of $x = 1$ error, the error rate is with 95% probability $\lambda < 2. \, 10^{-7}$ h^{-1}, as illustrated in Figure 4d). The target level of safety $\lambda_{SL} = 10^{-7}$ h^{-1} is thus not fulfilled on basis of the 95% quantile.

Influence of Error Dependence on Multi-Sensor Based Machine Vision

While the previous two sections dealt with the reliability assessment of an individual sensor, this section studies the reliability of a multi-sensor system where we model the sensor data fusion with a majority voting scheme. The multi-sensor system consists of three identical redundant sensors. First it is assumed that all sensors comply with the target level of safety such that each individual sensor has a perception error rate of exactly

FIGURE 4   a) Posterior PDF $f(\lambda|x,t)$ of the error rate when observing $x = 0$ errors in time $t = 1.92 \cdot 10^7$ h and b) corresponding CDF. c) Posterior PDF $f(\lambda|x,t)$ of the rate when observing $x = 1$ error in time $t = 1.92 \cdot 10^7$ h and d) corresponding CDF. The grey arrows indicate the 95 % quantiles of the error rate λ.

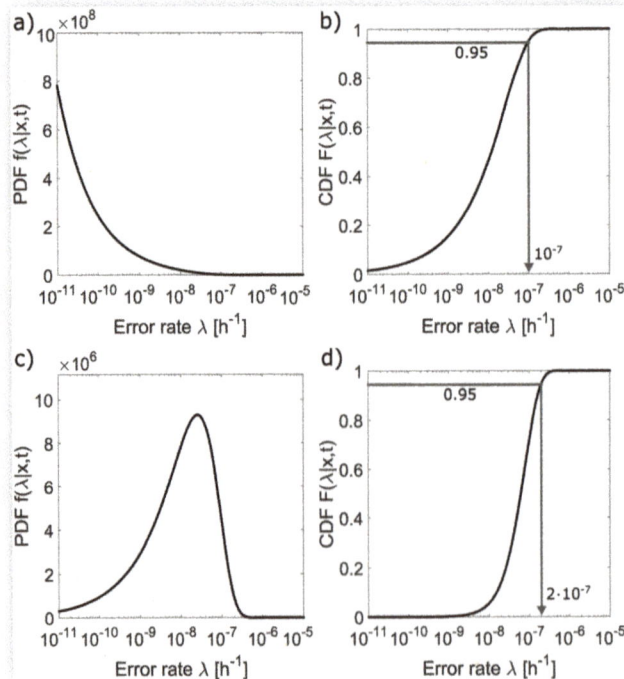

$\lambda = \lambda_{SL} = 10^{-7}$ h^{-1}. However, the correlation coefficient of error occurrence ρ between the three sensors is unknown. What is now the error rate λ_{system} of the multi-sensor based machine vision? The answer to this question can be obtained with Eqs. (31) to (33) for the beta-binomial distribution and with Eqs. (A1), (A2) and Eq. (33) for the Gupta and Tao model [48].

λ_{system} is illustrated in Figure 5 in function of the correlation coefficient ρ in semi- and in double-logarithmic scale. The semi-logarithmic plot in Figure 5 shows that the Gupta and Tao model (dashed line) in this specific case cannot be utilized for $\rho > 0.5$ and deviates from the beta-binomial distribution starting around $\rho = 0.05$. In the range of $\rho > 0.01$, we therefore utilize the beta-binomial model that converges against the individual sensors' perception error rate $\lambda = 10^{-7}$ h^{-1} with full dependence ($\rho \to 1$). This is the intuitive solution of a fully dependent redundant system. The beta-binomial model (solid line) cannot be evaluated for $\rho < 0.006$ due to numerical reasons. Thus, in the range $\rho < 0.01$ the Gupta-Tao model is utilized, which converges against the solution of the independent binomial CDF Eq. (26) with $\rho \to 0$. With independence, the system's error rate is $\lambda_{system} = 4.2 \cdot 10^{-19}$ h^{-1}.

In Figure 5 it is visible that for $\rho < 0.1$ the system's error rate λ_{system} strongly depends on the correlation coefficient. The reason for this sensitivity is explained with the interpretation of the correlation coefficient given through Eq. (29) and (30). For instance if $\rho \approx Pr\,(U_s = 1|\,U_q = 1) = 10^{-5}$, then the conditional probability of perception error occurrence in sensor s - given an error has occurred in sensor q - is $7.2 \cdot 10^6$ times larger than in the independent case. As can be seen in Figure 5b, this leads to a system error rate $\lambda_{system} \approx 3 \cdot 10^{-12}$ h^{-1}, substantially larger than with independent component errors.

In the previous sections of this case study, the test drive effort was derived such that an individual sensor complies with $\lambda_{SL} = 10^{-7}$ h^{-1}. However, the relationship between the system's error rate λ_{system} and correlation among the sensors has implications on the test effort, when the target level of safety is set on the system level. The two important questions then are: How large does the perception error rate of an individual sensor has to be and how much test drive effort is necessary at the individual sensor level, such that

FIGURE 5 a) Perception error rate λ_{system} of a redundant multi-sensor based machine vision system in dependence of the correlation coefficient ρ. The system consists of three identical sensors with $\lambda = \lambda_{SL} = 10^{-7}$ h^{-1} each. The solid line represents the beta-binomial model and the dashed line the model presented in [48] (see appendix). b) The same plot in double logarithmic scale.

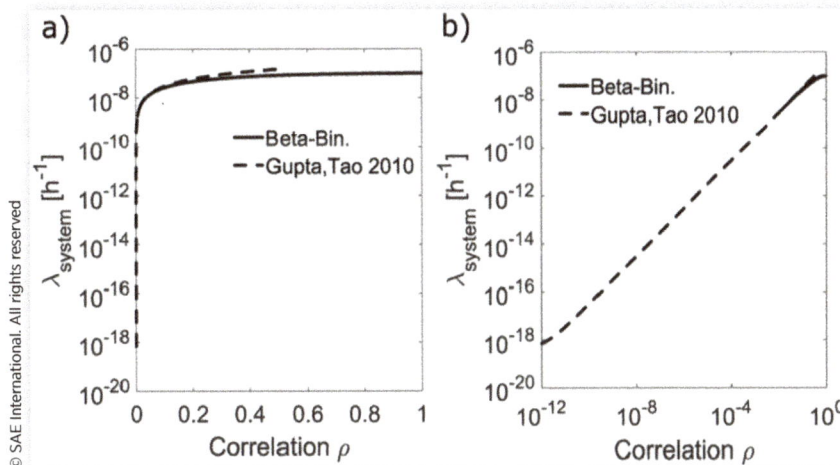

a) Necessary target level of safety for the individual sensors $\lambda_{individual,\,SL}$ in a redundant multi-sensor system (3 identical sensors) such that the system complies with $\lambda_{system} \leq \lambda_{SL} = 10^{-7}\,\mathrm{h}^{-1}$. b) Corresponding test drive effort t for $\Pr(\lambda \leq \lambda_{individual,\,SL}|x,t) = 0.95$ when no error is accepted in t. Both a) and b) are in function of the correlation coefficient ρ of error occurrence among the different sensors.

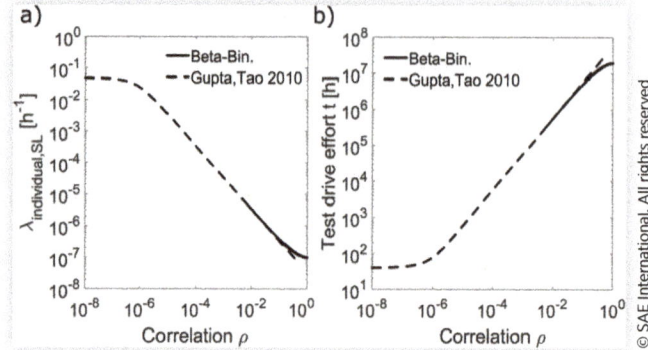

the system complies with the target level of safety $\lambda_{system} \leq \lambda_{SL} = 10^{-7}\mathrm{h}^{-1}$? The solution to these questions is illustrated in Figure 6 in function of the error correlation among different sensors.

For clarity, the target level of safety for the individual sensors in a redundant multi-sensor system that lead to an overall system perception error rate of $\lambda_{SL} = 10^{-7}\,\mathrm{h}^{-1}$ is denoted with $\lambda_{individual,\,SL}$ and is shown in Figure 6a. Figure 6b shows the corresponding test drive effort which is necessary to demonstrate that the target level of safety of an individual sensor $\lambda_{individual,\,SL}$ is complied with $\Pr(\lambda \leq \lambda_{individula,\,SL}|x,t) = 0.95$. If all sensors are independent of each other, the individual safety target is as low as $\lambda_{individual,\,SL} = 0.05\,\mathrm{h}^{-1}$ only requiring a test drive effort of $t = 40$ h to demonstrate that the system complies with the target level of safety, i.e. $\Pr(\lambda \leq \lambda_{individual,\,SL}|x,t) = \Pr(\lambda_{system} \leq \lambda_{SL}) = 0.95$. In contrast, if all sensors are fully dependent, the safety target of the individual sensors reduces to $\lambda_{individual,\,SL} = \lambda_{SL} = 10^{-7}\mathrm{h}^{-1}$ with a test drive effort of $t = 1.92 \cdot 10^{7}$ h. Obviously this is the same test drive effort as when implementing only a single sensor. With a correlation coefficient of $\rho \leq 10^{-4}$ the test drive effort is $t \leq 5.9 \cdot 10^{3}\mathrm{h}$.

Discussion

The interest is in the test drive effort that allows an empirical compliance demonstration of a sensors' environment perception with the (for demonstrative purposes selected) target level of safety $\lambda_{SL} = 10^{-7}\,\mathrm{h}^{-1}$. For individual sensors, this test drive effort is found to be in the order of at least $t = 1.92 \cdot 10^{7}\mathrm{h}$, which appears infeasible in most practical contexts. Similar conclusions have already been drawn in [14], in which the test drive effort to demonstrate an automated system's safety is derived with the Null Hypothesis Significance Testing (NHST). This indicates that the empirical demonstration (i.e. test drives in real driving situations) might not be the way to show that a sensor's perception is sufficiently safe when the target level of safety is strict.

In case of less restrictive safety requirements however, the here presented Bayesian approach allows to estimate test efforts. It has two main advantages over NHST: 1) it provides results that are easy to interpret, and 2) it is more flexible in extending the model, in particular to a hierarchical model that addresses changing environmental factors, and to a multi-sensor system. Figure 4 shows how the presented Bayesian approach captures the full uncertainty in estimating the perception error rate λ, which is more intuitively understood than statements on the statistical significance of a hypothesis about λ. The graphical illustrations in Figure 4 and Figure 1 underline that the Bayesian methodology can readily be interpreted as a probability, while the p-value and significance level α of NHST are more difficult to understand (and are not understood by most engineers). Due to its intuitive and easy interpretation, the communication with decision makers benefits from the Bayesian approach: Given that the statistical model and assumptions represent the problem adequately well, the Bayesian test methodology results in the probability $\Pr(\lambda < \lambda_{SL} | x, t)$ that for a specific set of observations the target level of safety is complied with.

In case certain aspects of the environment perception such as object localization are based on multiple redundant sensors, the redundancy should be considered in the reliability assessment of environment perception. As in every redundant system, redundancy can drastically increase the system reliability if sensors perform independently. However, Figure 5 should serve as a warning not to assume independence light handed, as the true error rate of the perception system might then be underestimated. For the investigated system, a seemingly small correlation of $\rho = 10^{-5}$ increases the system's perception error rate by a factor of $7.1 \cdot 10^6$ compared to the case of independence ($\rho = 0$)!

It is illustrated in Figure 6 that the overall target level of safety λ_{SL} of the machine vision may be more easily demonstrated in a redundant multi-sensor system than for an individual sensor, as long as errors dependence among the different sensors is small. In the extreme case of error independence at different sensors ($\rho = 0$), a system consisting of three sensors would require a test effort of only $t = 40$ h to comply with $\lambda_{SL} = 10^{-7}$ h$^{-1}\lambda$. This illustrates how a redundant sensor system may offer the opportunity to demonstrate the safety of the environment perception with economical feasible effort in an empirical way, i.e. test drives with individual sensors in real driving situations. This result may be especially relevant for the future when the costs of environment sensors decrease. For instance, the reliability of the environment perception could be more easily and with lower costs ensured by multiple (independent) mid-class sensors than with one high-end sensor. However, in order to evaluate whether this is a valid alternative, one would need to know the correlation ρ of error occurrence among different sensors. The problem therefore shifts to demonstrating a low correlation ρ, which likely will require a larger test effort. This study should therefore be extended to determine the test effort necessary to determine a sufficiently low correlation between the sensors.

It is important to note that the presented approach does not model the sensor data fusion in its full complexity. Instead of analyzing real fusion algorithms, the results are here derived by simplifying the sensor data fusion with a majority decision. While majority voting is a valid method to increase a system's reliability [12], it is doubtful whether this method finds widespread application in practice. It is likely that modern fusion algorithms [43] will outperform a majority voting system, therefore the presented results may be regarded as conservative.

The presented study combines different types of perception errors into the single metric "error rate λ". In reality, different types of perception errors include false-positive and false-negative object detections, errors in physical measurement quantities and object classification errors. The different types of errors occur with varying probability and may not be equally safety-relevant. It is here entirely left to the analyst to decide what comprises a safety-relevant perception error.

Another limitation is found in the treatment of the temporal variability of error occurrence, induced through different physical influencing conditions. In this contribution we do not answer the question of how to assess which factors with influence on sensor performance to consider, this has to be decided by experts and preliminary test results. Also, as stated before, the probability of the different influencing factors such as weather is dependent on the geographical region. Future work should optimize the presented work to account for this aspect and examine how to include different environment conditions in higher detail.

Conclusions

A Bayesian methodology for empirical reliability assessments of sensor based environment perception is presented as an alternative to the commonly applied Null Hypothesis Significance Testing (NHST). It allows to estimate the necessary test drive effort to demonstrate the perception reliability of environment sensors, including dependent errors and time variable error probabilities. Furthermore, a solution to assess the reliability of a dependent redundant multi-sensor system is given.

Applying the methodology in a case study shows that the empirical test drive effort may be unfeasibly large when the target level of safety is low. When working with a multi-sensor system in which the individual sensors are nearly independent of each other, the system's perception reliability is considerably higher than when utilizing a single sensor. This fact opens up the possibility of validating the perception reliability empirically with feasible test drive effort, when one is able to show that multiple sensors have a small error dependency. The verification of a small error dependency itself is however expected to require additional test drive efforts. It remains to be studied what the necessary test setup and effort is for this purpose. Simplifications of the problem's complexity involve the treatment of different types of perception errors, the representation of the sensor data fusion with a majority voting scheme and in approximating the time dependent performance of the perception induced through various physical influencing factors such as the weather.

Contact Information

Mario Berk, Ph.D. student at the Engineering Risk Analysis Group Technical University of Munich in cooperation with AUDI AG, INI.
TUM
mario.berk@tum.de

Daniel Straub
Professor of Engineering Risk Analysis, Technical University of Munich
straub@tum.de

References

1. Beiker, S., Deployment Scenarios for Vehicles with Higher-Order Automation, Maurer, M., Gerdes, J.C., Lenz, B., and Winner, H. (eds.), *Autonomous Driving* (Berlin, Heidelberg: Springer Berlin Heidelberg, 2016), ISBN 978-3-662-48845-4:193-211.

2. Google, "Google Self-Driving Car Project," accessed February 25, 2016, http://static.googleusercontent.com/media/www.google.com/en/us/selfdrivingcar/.

3. Audi, "Mission Accomplished: Audi A7 Piloted Driving Car Completes 550-Mile Automated Test Drive," accessed February 25, 2016, https://www.audiusa.com/newsroom/news/press-releases/2015/01/550-mile-piloted-drive-from-silicon-valley-to-las-vegas.

4. Pink, O., Becker, J., and Kammel, S., "Automated Driving on Public Roads: Experiences in Real Traffic," *IT-Information Technology* 57, no. 4 (2015), doi:10.1515/itit-2015-0010.

5. Broggi, A., Debattisti, S., Grisleri, P., and Panciroli, M., "The Deeva Autonomous Vehicle Platform," *2015 IEEE Intelligent Vehicles Symposium (IV)*, Seoul, South Korea, 2015, 692-699.

6. Aeberhard, M., Rauch, S., Bahram, M., Tanzmeister, G. et al., "Experience, Results and Lessons Learned from Automated Driving on Germany's Highways," *IEEE Intell. Transport. Syst. Mag.* 7, no. 1 (2015): 42-57, doi:10.1109/MITS.2014.2360306.

7. Franke, U., Pfeiffer, D., Rabe, C., Knoeppel, C. et al., "Making Bertha See," *2013 IEEE International Conference on Computer Vision Workshops (ICCVW)*, Sydney, Australia, 2013, 214-221.

8. Kammel, S., Ziegler, J., Pitzer, B., Werling, M. et al., "Team AnnieWAY's Autonomous System for the 2007 DARPA Urban Challenge," *J. Field Robotics* 25, no. 9 (2008): 615-639, doi:10.1002/rob.20252.

9. Blevis, B., "Losses Due to Rain on Radomes and Antenna Reflecting Surfaces," *IEEE Trans. Antennas Propagat.* 13, no. 1 (1965): 175-176, doi:10.1109/TAP.1965.1138384.

10. Ishimaru, A., "Wave Propagation and Scattering in Random Media and Rough Surfaces," *Proc. IEEE* 79, no. 10 (1991): 1359-1366, doi:10.1109/5.104210.

11. Rasshofer, R.H., Spies, M., and Spies, H., "Influences of Weather Phenomena on Automotive Laser Radar Systems," *Adv. Radio Sci.* 9 (2011): 49-60, doi:10.5194/ars-9-49-2011.

12. Weitzel, A., Winner, H., Peng, C., Geyer, S. et al., Absicherungsstrategien für Fahrerassistenzsysteme mit Umfeldwahrnehmung: [Bericht zum Forschungsprojekt FE 82.0546/2012], *Berichte der Bundesanstalt für Straßenwesen Fahrzeugtechnik* (Fachverl, NW: Bremen, 2014), vol. 98, ISBN 978-3-95606-118-9.

13. Wachenfeld, W. and Winner, H., The Release of Autonomous Vehicles, Maurer, M., Gerdes, J.C., Lenz, B., and Winner, H. (eds.), *Autonomous Driving* (Berlin Heidelberg: Springer Berlin, Heidelberg, 2016), ISBN 978-3-662-48845-4:425-449.

14. Winner, H., ADAS, Quo Vadis?, Winner, H., Hakuli, S., Lotz, F., and Singer, C. (eds.), *Handbook of Driver Assistance Systems* (Cham: Springer International Publishing, 2016), ISBN 978-3-319-12351-6:1557-1584.

15. Bock, F., Siegl, S., and German, R., "Mathematical Test Effort Estimation for Dependability Assessment of Sensor-based Driver Assistance Systems," *2016 42st Euromicro Conference on Software Engineering and Advanced Applications (SEAA)*, Limassol, Cyprus, 2016.

16. Greenland, S., Senn, S.J., Rothman, K.J., Carlin, J.B. et al., "Statistical Tests, P Values, Confidence Intervals, and Power: A Guide to Misinterpretations," *European Journal of Epidemiology* 31, no. 4 (2016): 337-350, doi:10.1007/s10654-016-0149-3.

17. Cohen, J., "The Earth is Round (p < .05)," *American Psychologist* 49, no. 12 (1994): 997-1003, doi:10.1037/0003-066X.49.12.997.

18. Nuzzo, R., "Scientific Method: Statistical Errors," *Nature* 506 (2014): 150-152, doi:10.1038/506150a.

CHAPTER 8

19. Baker, M., "Statisticians Issue Warning over Misuse of P Values," *Nature* 531 (2016): 151, doi:10.1038/nature.2016.19503.

20. Wasserstein, R.L. and Lazar, N.A., "The ASA's Statement on p -Values: Context, Process, and Purpose," *The American Statistician* 70, no. 2 (2016): 129-133, doi:10.1080/00031305.2016.1154108.

21. Hamada, M.S., Wilson, A.G., Reese, C.S., and Martz, H.F., Bayesian Reliability, *Springer Series in Statistics*, 1st ed., (Springer-Verlag, s.l., 2008), ISBN 978-0-387-77950-8.

22. Rausand, M. and Høyland, A., System Reliability Theory: Models, Statistical Methods, and Applications, *Wiley Series in Probability and Statistics Applied Probability and Statistics*, 2nd ed., (Hoboken, NJ: Wiley-Interscience, 2004), ISBN 978-0-471-47133-2.

23. Fitzgerald, M., Martz, H.F., and Parker, R.L., "Bayesian Single-Level Binomial and Exponential Reliability Demonstration Test Plans," *Int. J. Rel. Qual. Saf. Eng.* 06, no. 02 (1999): 123-137, doi:10.1142/S0218539399000139.

24. Singh, H., Cortellessa, V., Cukic, B., Gunel, E. et al., "A Bayesian Approach to Reliability Prediction and Assessment of Component Based Systems," *12th International Symposium on Software Reliability Engineering, ISSRE 2001*, Hong Kong, China, November 27-30, 2001, 12-21.

25. Brender, D.M., "The Bayesian Assessment of System Availability: Advanced Applications and Techniques," *IEEE Trans. Rel.* R-17, no. 3 (1968): 138-147, doi:10.1109/TR.1968.5216927.

26. Martz, H.F. and Wailer, R.A., "Bayesian Reliability Analysis of Complex Series/Parallel Systems of Binomial Subsystems and Components," *Technometrics* 32, no. 4 (1990): 407-416.

27. Guida, M. and Pulcini, G., "Automotive Reliability Inference Based on Past Data and Technical Knowledge," *Reliability Engineering & System Safety* 76, no. 2 (2002): 129-137, doi:10.1016/S0951-8320(01)00132-6.

28. Gotzig, H. and Geduld, G., Automotive LIDAR, Winner, H., Hakuli, S., Lotz, F., and Singer, C. (eds.), *Handbook of Driver Assistance Systems* (Cham: Springer International Publishing, 2016), 405-430, ISBN 978-3-319-12351-6.

29. Winner, H., Automotive RADAR, Winner, H., Hakuli, S., Lotz, F., and Singer, C. (eds.), *Handbook of Driver Assistance Systems* (Cham: Springer International Publishing, 2016), 325-403, ISBN 978-3-319-12351-6.

30. Dietmayer, K., Predicting of Machine Perception for Automated Driving, Maurer, M., Gerdes, J.C., Lenz, B., and Winner, H. (eds.), *Autonomous Driving* (Berlin, Heidelberg: Springer Berlin Heidelberg, 2016), 407-424, ISBN 978-3-662-48845-4.

31. Kottegoda, N.T. and Rosso, R., *Applied Statistics for Civil and Environmental Engineers*, 2nd ed., (Oxford, UK [Malden, MA]: Blackwell Pub, 2008), ISBN 978-1-4051-7917-1.

32. Vidgen, B. and Yasseri, T., "P-Values: Misunderstood and Misused," *Front. Phys.* 4 (2016): e124, doi:10.3389/fphy.2016.00006.

33. Morey, R.D., Hoekstra, R., Rouder, J.N., Lee, M.D. et al., "The Fallacy of Placing Confidence in Confidence Intervals," *Psychonomic Bulletin & Review* 23, no. 1 (2016): 103-123, doi:10.3758/s13423-015-0947-8.

34. VanderPlas, J., "Frequentism and Bayesianism: A Python-Driven Primer," arXiv preprint arXiv:1411.5018, 2014.

35. Jaynes, E.T. and Kempthorne, O., Confidence Intervals vs Bayesian Intervals, Harper, W.L. and Hooker, C.A. (eds.), *Foundations of Probability Theory, Statistical Inference, and Statistical Theories of Science* (Dordrecht: Springer Netherlands, 1976), 175-257, ISBN 978-90-277-0619-5.

36. Gelman, A., "Induction and Deduction in Bayesian Data Analysis," *Rationality, Markets and Morals* 2 (2011): 67-78.

37. Darms, M., Data Fusion of Environment-Perception Sensors for ADAS, Winner, H., Hakuli, S., Lotz, F., and Singer, C. (eds.), *Handbook of Driver Assistance Systems* (Cham: Springer International Publishing, 2016), 549-566, ISBN 978-3-319-12351-6.

38. Winner, H., Fundamentals of Collision Protection Systems, Winner, H., Hakuli, S., Lotz, F., and Singer, C. (eds.), *Handbook of Driver Assistance Systems* (Cham: Springer International Publishing, 2016), 1149-1176, ISBN 978-3-319-12351-6.

39. Straub, D., *Lecture Notes in Engineering Risk Analysis* (2013).

40. Gelman, A., Carlin, J.B., Stern, H.S., and Rubin, D.B., Bayesian Data Analysis, *Texts in Statistical Science*, 2nd ed., (Boca Raton, FL: Chapman & Hall, 2004), ISBN 9781584883883.

41. Jeffreys, H., *Theory of Probability*, 3rd ed., (London: Oxford University Press, 1961).

42. Kass, R.E. and Wasserman, L., "The Selection of Prior Distributions by Formal Rules," *Journal of the American Statistical Association* 91, no. 435 (1996): 1343-1370.

43. Durrant-Whyte, H. and Henderson, T.C., Multisensor Data Fusion, Siciliano, B. and Khatib, O. (eds.), *Springer Handbook of Robotics* (Berlin, Heidelberg: Springer Berlin Heidelberg, 2008), 585-610, ISBN 978-3-540-23957-4.

44. Hisakado, M., Kitsukawa, K., and Mori, S., "Correlated Binomial Models and Correlation Structures," *Journal of Physics A: Mathematical and General* 39, no. 50 (2006): 15365.

45. Rosner, B., Beta-Binomial Distribution, Armitage, P., and Colton, T. (eds.), *Encyclopedia of Biostatistics* (Chichester, UK: John Wiley & Sons, Ltd, 2005), ISBN 047084907X.

46. Maharry, T.J., "Proportion Differences Usingthe Beta-Binomial Distribution," Dissertation, Oklahoma State University, Stillwater, Oklahoma, 2006.

47. Paul, S.R., Applications of the Beta Distribution, Gupta, A.K. and Nadarajah, S. (eds.), *Handbook of Beta Distribution and Its Applications* (CRC Press, 2004), 423-436.

48. Gupta, R.C. and Tao, H., "A Generalized Correlated Binomial Distribution with Application in Multiple Testing Problems," *Metrika* 71, no. 1 (2010): 59-77, doi:10.1007/s00184-008-0202-7.

49. Bahadur, R.R., "A Representation of the Joint Distribution of Responses to n Dichotomous Items," *Studies in Item Analysis and Prediction* 6 (1961): 158-168.

50. Kupper, L.L. and Haseman, J.K., "The Use of a Correlated Binomial Model for the Analysis of Certain Toxicological Experiments," *Biometrics* (1978): 69-76.

Appendix

To complement the results of the beta-binomial distribution, the correlated binomial distribution suggested in Gupta, Tao [48] is used. With this model the probability

$\Pr\left(\sum_{s=1}^{N} U_s = k\right)$ of exactly k-out-of-N sensors to commit a safety-relevant error, assuming each sensor has the same perception error probability p, is:

$$\Pr\left(\sum_{s=1}^{N} U_s = k\right) =$$
$$p \cdot \Pr\left(\sum_{s=1}^{N-1} U_s = k-1\right) + \left(1-p\right) \cdot \Pr\left(\sum_{s=1}^{N-1} U_s = k\right) + \rho \cdot \sum_{s=1}^{N-1} p(1-p) \cdot a_{N,k}^s \tag{A1}$$

Where ρ is the correlation coefficient quantifying the correlation of error occurrence between the sensors. It is assumed ρ is equal among all sensors $s = 1, \ldots, N$. The factor $a_{N,k}^S$ is defined as:

$$a_{N,k}^s = \begin{cases} 0, & if\ k < 0\ or\ k > N \\ a_{2,0}^1 = 1, a_{2,1}^1 = -2, a_{2,2}^1 = 1, & if\ N = 2, \quad s = 1 \\ p.a_{N-1,k-1}^{S-1} + (1-p)a_{N-1,k}^{S-1}, & if\ N > 2, \quad s = N-1 \\ p.a_{N-1,k-1}^{S} + (1-p)a_{N-1,k}^{S}, & if\ N > 2, \quad s = 1,2,\ldots,N-2 \end{cases} \tag{A2}$$

Inserting Eq. (A1) into Eq. (33) yields the probability of the majority vote based multi-sensor machine vision to fail p_f, including dependence according to the Gupta, Tao model.

Challenges in Autonomous Vehicle Testing and Validation

Philip Koopman
Carnegie Mellon University

Michael Wagner
Edge Case Research LLC

S oftware testing is all too often simply a bug hunt rather than a well-considered exercise in ensuring quality. A more methodical approach than a simple cycle of system-level test-fail-patch-test will be required to deploy safe autonomous vehicles at scale. The ISO 26262 development V process sets up a framework that ties each type of testing to a corresponding design or requirement document, but presents challenges when adapted to deal with the sorts of novel testing problems that face autonomous vehicles. This paper identifies five major challenge areas in testing according to the V model for autonomous vehicles: driver out of the loop, complex requirements, non-deterministic algorithms, inductive learning algorithms, and fail-operational systems. General solution approaches that seem promising across these different challenge areas include: phased deployment using successively relaxed operational scenarios, use of a monitor/actuator pair architecture to separate the most complex autonomy functions from simpler safety functions, and fault injection as a way to perform more efficient edge case testing. While significant challenges remain in safety-certifying the type of algorithms that provide high-level autonomy themselves, it seems within reach to instead architect the system and its accompanying design process to be able to employ existing software safety approaches.

CITATION: Koopman, P. and Wagner, M., "Challenges in Autonomous Vehicle Testing and Validation," *SAE Int. J. Trans. Safety* 4(1):2016, doi:10.4271/2016-01-0128.

Introduction

While self-driving cars have recently become a hot topic, the technology behind them has been evolving for decades, tracing back to the Automated Highway System project [1], and before. Since those early demonstrations, the technology has matured to the point that Advanced Driver Assistance Systems (ADAS) such as automatic lane keeping and smart cruise control are standard on a number of vehicles. Beyond that, there are numerous different fully autonomous vehicle projects in various stages of development, including extended on-road testing of multi-vehicle fleets.

If one believes pundits, full-scale fleets of autonomous vehicles (often called "self-driving cars") are just around the corner. However, as the traditional automotive industry knows well, there is a huge difference between building a few vehicles to run in reasonably benign conditions with professional safety drivers, and building a fleet of millions of vehicles that have to run in an unconstrained world. Some say that successful demonstrations and a few thousand km (or even a few hundred thousand km) of driving experience means that autonomous vehicle technology is essentially ready to be deployed at full scale. But, it is difficult to see how such testing alone would be enough to ensure adequate safety. Indeed, at least some developers seem to be doing more, but the question is how much more might be required, and how we can know that the resultant vehicles are sufficiently safe to deploy.

In this paper we explore some of the challenges that await developers who are attempting to qualify fully autonomous, NHTSA Level 4 [2] vehicles for large-scale deployment. Thus, we skip past potential semi-automated approaches to address systems in which the driver is not responsible at all for safe vehicle operation. We further limit scope to consider how such vehicles might be designed and validated within the ISO 26262 V framework. The reason for this constraint is that this is an acceptable practice for ensuring safety. It is a well-established safety principle that computer-based systems should be considered unsafe unless convincingly argued otherwise (i.e., safety must be shown, not assumed). Therefore, autonomous vehicles cannot be considered safe unless and until they are shown to conform or map to ISO 26262 or some other suitable, widely accepted software safety standard.

Infeasibility of Complete Testing

Vehicle-level testing won't be enough to ensure safety. It has long been known that it is infeasible to test systems thoroughly enough to ensure ultra-dependable system operation.

For example, consider a hypothetical fleet of one million vehicles operated one hour per day (i.e., 10^6 operational hours per day). If the safety target is to have about one catastrophic computing failure in this fleet every 1,000 days, then the safety goal is a mean time between catastrophic failures of 10^9 hours, which is comparable to aircraft permissible failure rates [3]. Note that this admits to the likelihood that several such catastrophic failures due to computer defects or malfunctions will happen during the life of the fleet of cars. However, such a goal might be justifiable if accompanied by a much larger reduction in catastrophic mishaps due to driver error compared to manually driven vehicles. (This is just an example failure rate. Arguments might be made for this rate to be higher or lower, but it has been selected as a defensible rate that illustrates some of the difficulties in achieving safety.)

In order to validate that the catastrophic failure rate of a vehicle fleet is in fact one per 10^9 hours, one must conduct at least 10^9 vehicle operational hours of testing (a billion hours) [4], and in fact must test several times longer, potentially repeating such tests multiple times to achieve statistical significance. Even this assumes that the testing environment is highly representative of real-world deployment, and that circumstances causing mishaps arrive in a random, independent manner. Building a fleet of physical vehicles big enough to run billions of hours in representative test environments without endangering the public seems impractical. Thus, alternate methods of validation are required, potentially including approaches such as simulation, formal proofs, fault injection, bootstrapping based on a steadily increasing fleet size, gaining field experience with component technology in non-critical roles, and human reviews. (Component level testing also plays a role, but it is still impractical to accumulate 10^9 hours of pre-deployment testing for a physical hardware device.) Things get even worse when one considers that testing is even more difficult for autonomy systems than for everyday software systems, as will be discussed below.

That having been said, for relatively non-critical computing systems it may be possible to use testing as a primary basis for validating an appropriate level of safety. This is because failures involving low severity and low exposure may be permissible at a higher occurrence rate than catastrophic failures. For example, if a failure of a particular type once every 1,000 hours is acceptable (because such failures result in a minimal-cost incident or slight disruption), then validation of that failure rate could be credibly achievable by testing for several thousand hours. This is not to say that all software quality process can be abandoned for such systems, but rather that a suitable testing and failure-monitoring strategy might make it possible to validate that a component with suitable quality has actually attained an acceptably low failure rate if the mean-time-between-failure requirement is relatively lenient.

The V Model as a Starting Point

Because system-level testing can't do the job, more is required. And that is precisely the point of having a more robust development framework for creating safety critical software.

The "V" software development model has been applicable to vehicles for a long time. It was one of the development reference models incorporated into the MISRA Guidelines more than 20 years ago [5, 6]. More recently, it has been promoted to be the reference model that forms the basis of ISO 26262 [7].

In general, the V model (Figure 1) represents a methodical process of creation followed by verification and validation. The left side of the V works its way from requirements through design to implementation. At each step it is typical for the system to be broken into subsystems that are treated in parallel (e.g., there is one set of system requirements, but separate designs for each subsystem). The right side of the V iteratively verifies and validates larger and larger chunks of the system as it climbs back up from small components to a system-level assessment. While ISO 26262 has a detailed elaboration of this model, and much more, we keep things generic so as to discuss the high level ideas.

Although ISO 26262 and its V framework generally reflect accepted practices for ensuring automotive safety, fully autonomous vehicles present unique challenges in mapping the technical aspects of the vehicle to the V approach.

Driver Out of the Loop

Perhaps the most obvious challenge in a fully autonomous vehicle is that the whole point is for the driver to no longer be actually driving the vehicle. That means that, by definition, the driver can no longer be counted on to provide control inputs to the vehicle during operation [2].

CHAPTER 9

FIGURE 1 A generic V model.

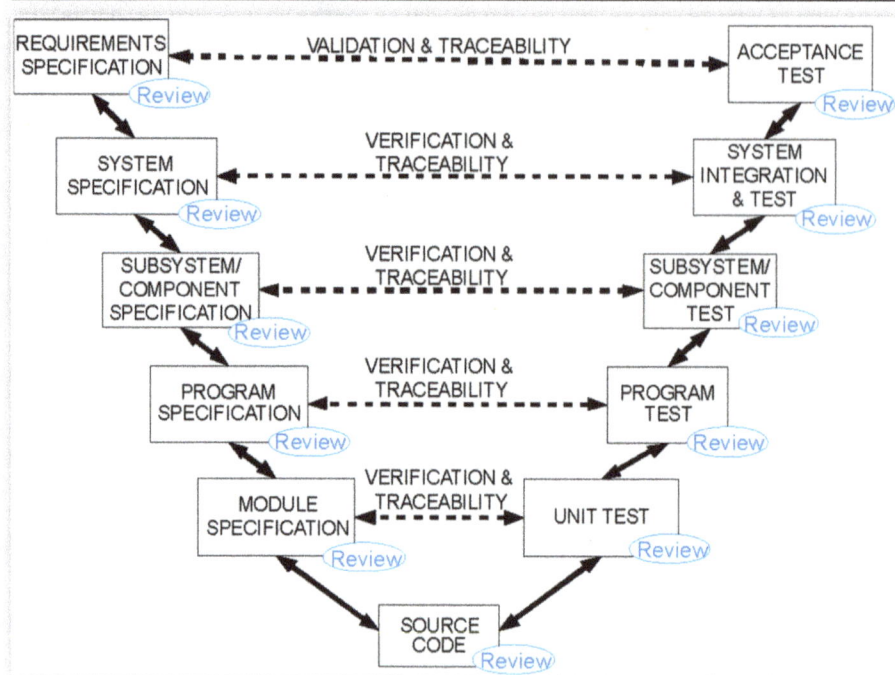

Controllability Challenges

Typical automotive safety arguments for low-integrity devices can hinge upon the ability of a human driver to exert control. For example, with an Advanced Driver Assistance System (ADAS), if a software fault causes a potentially dangerous situation, the driver might be expected to over-ride that software function and recover to a safe state. Drivers are also expected to recover from significant vehicle mechanical failures such as tire blow-outs. In other words, in human-driven vehicles the driver is responsible for taking the right corrective action. Situations in which the driver does not have an ability to take corrective action are said to lack controllability, and thus must be designed to a higher Automotive Safety Integrity Level, or ASIL [8].

With a fully autonomous vehicle, the driver can't be counted on to handle exceptional situations. Rather, the computer system must assume that role as the primary exception handler for faults, malfunctions, and beyond-specified operating conditions. Putting the computer in charge of exception handling seems likely to dramatically increase automation complexity compared to ADAS systems. Combinations of ADAS systems such as lane-keeping and smart cruise control seem tantalizingly close to fully autonomous operation. However, a fully autonomous vehicle must have significant additional complexity to deal with all the ways things might go wrong because there is no driver to grab the wheel and hit the brakes when something goes awry.

Autonomy Architecture Approaches

In the context of ISO 26262, putting the computer in charge suggests one of two strategies for assessing risk. One strategy is that the controllability portion of risk evaluation [8] should be set to "C3 Difficult to control or uncontrollable." This might be a viable option if the severity and exposure are very low, and thus a low ASIL can be assigned.

However, in cases that have moderate or high severity and exposure, the system must be designed to a high Automotive Safety Integrity Level (ASIL). (Some might argue that there should be an even higher controllability classification C4 because of the potential of an automation system to take proactively dangerous positive actions rather than simply failing to deliver a safety function. But we assume here that the existing C3 suffices.)

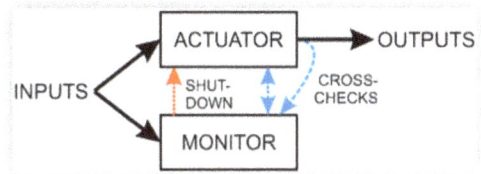

Another way to handle a potentially high-ASIL autonomy function is to use ASIL decomposition [9] via a combination of a monitor/ actuator architecture and redundancy. A monitor/actuator architecture is one in which the primary functions are performed by one module (the actuator), and a paired module (the monitor) performs an acceptance test [5, 10] or other behavioral validation. If the actuator misbehaves, the monitor shuts the entire function down (both modules), resulting in a fail-silent system (i.e., any failure results in a silent component, sometimes also known as fail-stop, or fail-safe).

If the monitor/actuator pair (Figure 2) is designed properly, the actuator can be designed to a low ASIL so long as the monitor has a sufficiently high ASIL and detects all possible faults in the monitor. (There is also a requirement to detect latent faults in the monitor to avoid a broken monitor failing to detect an actuator fault.) This architectural pattern can be especially advantageous if the monitor can be made substantially simpler than the actuator, reducing the size of the high-ASIL monitor, and permitting the majority of the functional complexity to be placed into a lower-ASIL actuator.

Both the strength and weakness of a monitor/actuator pair is that it creates a fail-silent building block (i.e., one that shuts down if there is a fault). The use of heterogeneous redundancy (two modules: the monitor and the actuator) is intended to prevent a malfunctioning actuator from issuing dangerous commands. However, it also causes loss of the actuator function if something goes wrong, which is a problem for a function that must fail operational, such as steering in a moving vehicle.

At the very least, providing fail operational behavior requires even more redundancy (more than one monitor/actuator pair), and very likely design diversity so that common-mode software design failures do not cause a systemic failure. This is important to avoid situations such as the loss of Arianne 5 Flight 501, which was caused by both a primary and a backup system that failed the same way due to experiencing the same un-handled exceptional (unanticipated by the component design) operating condition [11].

It should be noted that achieving diversity is not necessarily simple, due to issues such as vulnerability to defects in the same set of high-level requirements used to implement the diverse components (e.g., [12]). However this is a situation that is also true for nonautonomous software. It should also be noted that a monitor/actuator pair's fail-silent requirement is based on an assumption of failure independence, but again this is also true of non-autonomous systems.

A key high level point is that regardless of the approach, it seems likely that there will need to be a way to detect when autonomy functions are not working properly (whether due to hardware faults, software faults, or requirements defects), and to somehow bring the system to a safe state when such faults are detected via a fail-operational degraded mode autonomy capability.

Complex Requirements

An essential characteristic of the V model of development is that the right side of the V provides a traceable way to check how the left side turned out (verification and validation).

However, this notion of checking is predicated on an assumption that the requirements are actually known, are correct, complete, and unambiguously specified. That assumption presents challenges for autonomous vehicles.

Requirements Challenges

As mentioned earlier, removing the driver from the control system means that software has to handle exceptions, including weather, environmental hazards, and equipment failures. There are likely to be very many different types of these, from bad weather (flooding, fog, snow, smoke, tornados), to traffic rule violations (wrong-direction cars on a divided highway, other drivers running red lights, stolen traffic signs), to local driving conventions (parking chairs, the "Pittsburgh Left" [13]), to animal hazards (deer, armadillos, and the occasional plague of locusts).

Anyone who has driven for a long time is likely to have stories to tell of freak events they've seen on the road. A large fleet of vehicles will, in aggregate, be likely to experience all such types of events, and perhaps more. Worse still is that combinations of adverse events and driving conditions can occur that are simply too numerous to enumerate in a classical written requirements specification. Perhaps not all such extremely rare combinations need to be covered if results are likely to be innocuous, but the requirements should be clear about what is within the scope of system design, as well as what is not. Thus, it seems unlikely that a classical V process that starts with a document that enumerates all system requirements will be scalable to autonomous vehicle exception handling in a rigorous way, at least in the immediate future.

Operational Concept Approaches

One way to manage the complexity of requirements is to constrain operational concepts and engage in a phased expansion of requirements. This is already being done by developers who might concentrate on-road testing in particular geographic regions (for example only performing daytime driving on divided highways in Silicon Valley, which has limited precipitation and little freezing weather). However, the idea of employing an operational concept can be scaled in many directions.

Examples of axes that can be exploited for limiting operational concepts include:

- Road access: limited access highways, HOV lanes, rural roads, suburbs, closed campuses, urban streets, etc.

- Visibility: day, night, fog, haze, smoke, rain, snow, etc.

- Vehicular environment: self-parking in a closed garage with no other cars moving, autonomous-only lanes, marker transponders on non-autonomous vehicles, etc.

- External environment: infrastructure support, pre-mapped roads, convoying with human-driven cars

- Speed: lower speeds potentially lead to lower consequences of a failure and larger recovery margins

While there are still a great many combinations of the above degrees of freedom (and more that can no doubt be imagined), the purpose of selecting from possible operational concepts is not to increase complexity, but rather to reduce it. Mitigation of requirement complexity can be achieved via only enabling autonomy in a certain limited set of situations for which requirements are fully understood (and ensuring that the recognition of those valid operational conditions is correct).

Limiting operational concepts therefore becomes a bootstrapping strategy for deploying successively more sophisticated technical capabilities in a progressively more complex operational context (e.g., [14, 15]). Once confidence is gained that requirements for a particular operational concept are well understood, additional similar operational concepts can be added over time to expand the envelope of allowable automation scenarios. This will not entirely eliminate the issue of complex requirements, but it can help mitigate the combinatorial explosion of requirements and exceptions that would otherwise occur.

Safety Requirements and Invariants

Even with the use of restricted operational concepts, it seems likely that it will be impractical to use a traditional safety-related requirements approach. Such an approach more or less proceeds as follows. First the functional requirements are created. Then the requirements that are safety-relevant are annotated after some risk assessment process has been performed. Then, these safety-relevant requirements are allocated to safety critical subsystems. Then, safety critical subsystems are designed to satisfy allocated requirements. Finally, unanticipated emergent subsystem interactions are identified and mitigated via repeating the cycle.

Annotation of safety-critical requirements can be impractical for autonomy applications for at least two reasons. One reason is that many requirements might be only partially safety related, and are inextricably entwined with functional performance. For example, the many conditions for operating a parking brake when the car is moving could be a starting set of requirements. However, only some aspects of those requirements are actually safety critical, and those aspects are largely emergent effects of the interaction of the other functions. In the case of the parking brake, a deceleration profile when the parking brake is applied at speed is one of the desired functions, and is likely to be described by numerous functional requirements. But, simplifying, the only safety critical aspect in the deceleration mode might be that the emergent interaction of the other requirements must avoid locking up the wheels during the deceleration process.

The second reason that annotation of requirements to identify safety-relevant requirements may fail is that this may not even be possible when machine learning techniques are used. That is because the requirements, such as they are, take the form of a set of training data that enumerates a set of input values and correct system outputs. These tend not to be in the form of traditional requirements, and therefore require a different approach to requirements management and validation. (See the section on machine learning later in this paper).

Rather than attempting to allocate functional requirements among safety and non-safety subsystems, it can be helpful to create a separate, parallel set of requirements that are strictly safety related [16]. These requirements tend to be in the form of invariants that specify system states that are required for safety (both things that must be true to be safe, and things that must be false to be safe). This approach can disentangle issues of performance and optimization ("What is the shortest traveling path?" or "What is the speed for optimal fuel consumption?") from those of safety ("Are we going to hit anything?").

Using this approach would divide the set of requirements into two parts for the V model. The first set of requirements would be a set of non-safety-related functional requirements, which might be in traditional format or an untraditional format such as a machine learning training set. However, by definition those potentially nontraditional requirements are not safety-related, so it might be acceptable if traceability and validation have ample but imperfect coverage.

The second set of requirements would be a set of purely safety requirements that completely and unambiguously define what "safe" means for the system, relatively independent of the details of optimal system behavior. Such requirements can take the form of safe operating envelopes for different operational modes, with the system free to optimize its performance within the operating envelope [17]. It is clear that such envelopes can be used in at least some situations (e.g., enforcing a speed limit or a setting a minimum following distance). This concept promises to be rather general, but proving that remains future work.

A compelling reason to adopt a set of safety requirements that is orthogonal to functional requirements is that such an approach cleanly maps onto monitor/actuator architectures. Functional requirements can be allocated to a low-ASIL actuator functional block, while safety requirements can be allocated to a high-ASIL monitor. This idea has been used informally for many years as part of the monitor/actuator design pattern. We are proposing that this approach be elevated to a primary strategy for architecting autonomous vehicle designs, requirements, and safety cases rather than being relegated to a detailed implementation redundancy strategy.

Non-Deterministic and Statistical Algorithms

Some of the technologies used in autonomous vehicles are inherently statistical in nature. In general, they tend to be non-deterministic (non-repeatable), and may give answers that are only correct to some probability - if a probability can be assigned at all. Validating such systems presents challenges not typically found in more deterministic, conventional automotive control systems.

Challenges of Stochastic Systems

Non-deterministic computations include algorithms such as planners that might work by ranking the results of numerous randomly selected candidates (e.g., probabilistic roadmap planners [18]). Because the core operation of the algorithm is based on random generation of candidates, it is difficult to reproduce. While techniques such as using a reproducible pseudo-random number stream in unit test can be helpful, it may be impractical to create completely deterministic behavior in an integrated system, especially if small changes in initial conditions lead to diverging system behaviors. This means that every vehicle-level test could potentially result in a different outcome despite attempts to exercise nominally identical test cases.

Successful perception algorithms also tend to be probabilistic. For example, the evidence grid framework [19] accumulates diffuse evidence from individual, uncertain sensor readings into increasingly confident and detailed maps of a robot's surroundings. This approach yields a *probability* that an object is present, but never complete confidence. Furthermore, these algorithms are based on prior models of sensor physics (e.g., multipath returns) and noise (e.g., Gaussian noise on LIDAR-reported ranges) which are themselves probabilistic and sensitive to small changes in environmental conditions.

Beyond modeling the geometry of surroundings, other algorithms extract labels from perceived data. Prominent examples of these include pedestrian detection [20]. Such systems can exhibit potentially unpredicted failure modes even with largely noise-free data. For example, vision systems might have trouble disambiguating color variations due to shadows, and experience difficulties determining object positions in the presences of large reflective surfaces. (In all fairness, these present challenges for humans as well.) Moreover, any classification process exhibits a tradeoff between false negatives and false positives, with fewer of one necessarily incurring more of the other. The testing implications of this are that such algorithms won't "work" 100% of the time, and that depending on construction they might report a particular situation as being "true" when it is only a moderately high probability of that situation actually being true.

Non-Determinism in Testing

Handling non-determinism in testing is difficult for at least two reasons. The first is that it can be difficult to exercise a particular specific edge-case situation. This is because the system might behave in a way that activates that edge case only if it receives a very specific sequence of inputs from the world. Due to factors discussed earlier, such as the potentially dramatic differences in planner response to small changes of inputs, it can be difficult to contrive a situation in which the world will reliably offer up just the right conditions to run a particular desired test case.

As a simple example, a vehicle might prefer to drive a more circuitous route on a wide roadway rather than a shortcut through a narrow alley. To evaluate the performance navigating the narrow alley, testers would need to contrive a situation that makes the wide roadway unappealing to the planner. But, doing this requires additional attention to test planning, and perhaps (manually) moving the vehicle into a situation it would not normally enter to force the desired response. Testing the vehicle's ability to consistently choose the better of two almost equally unattractive paths without vacillating might be even more difficult.

A second difficulty with non-determinism in testing is that it can be difficult to evaluate whether test results are correct or not, because there is no unique correct system behavior for a given test case. Thus, correctness criteria are likely to have to take a form similar to the safety envelopes previously discussed, in which a test passes if the end system state is within an acceptable "test pass" envelope. In general, multiple tests might be required to build confidence that the system will always end up in the test pass envelope.

Probabilistic system behaviors present a similar challenge to validation, because passing a test once does not mean that the test will be passed every time. In fact, with a probabilistic behavior it might be expected that at least some types of tests will fail some fraction of the time. Therefore, testing might not be oriented toward determining if behaviors are correct, but rather to validating that the statistical characteristics of the behavior are accurately specified (e.g., that the false-negative detection rate is no greater than the rate assumed in an accompanying safety argument). This is likely to take a great many more tests than simple functional validation, especially if the behavior in question is safety critical and expected to have an extremely low failure rate.

Achieving extremely high performance from a probabilistic system is likely to require multiple subsystems that in composite are assumed to provide a low aggregate failure rate due to having completely independent failures. For example, a composite radar and vision system might be combined to assure no missed obstacles to within some extremely low probability. This approach applies not only to sensing modalities, but also to other diverse algorithmic schemes in planning and execution. If such an approach is successful, it might well be that the resulting probability of failure is so low that testing to verify the composite performance is infeasible. For example, if obstacles must be missed by both systems once in a billion detections, then billions of representative tests must be run to validate this performance.

Validating very low failure rates for composite diverse algorithms might be attempted by separately validating the more frequent permissible failure rates of each algorithm in isolation. But that is insufficient. One must also validate the assumption of independence between failures, which might well have to be based on analysis in addition to testing.

Machine Learning Systems

Proper behavior for autonomous vehicles is only possible if a complex series of perception and control decisions are made correctly. Achieving this usually requires proper tuning of parameters, including everything from a calibrated model of each camera lens to the well-tuned weighting of the risks of swerving versus stopping to avoid obstacles on a highway. The challenge here is to find the calibration model or the ratio of weights such that some error function is minimized. In recent years, most robotics applications have turned to machine learning to do this [21, 22], because the complexities of the multi-dimensional optimization are such that manual effort is unlikely to yield desired levels of performance.

The details of approaches to machine learning are many, e.g., the use of learning from demonstration, active learning, and supervised vs. unsupervised approaches. However, all such approaches involve inductive learning, in which training examples are used to derive a model.

For example, consider the case of detecting pedestrians in monocular images. Using a large training set of images, a classifier can learn a decision rule that minimizes the probability that pedestrians are detected in a separate validation set of images. For our purposes, an essential element is that the training set is effectively the set of requirements for the system, and the rules are the resultant system design. (The machine learning algorithm itself and the classifier algorithm are both more amendable to traditional validation techniques. However, these are general-purpose software "engines" and the ultimate system behaviors are determined by what training data is used for learning.)

One could attempt to skirt the issue of training set data forming de facto requirements by instead creating a set of requirements for collecting the training data. But this ends up simply pushing the same challenge up one level of abstraction. The requirements are not in the typical V format of a set of functional requirements for the system itself, but rather in the form of a set of training data or a plan for collecting the set of training data. How to validate training data is an open question that might be addressed by some combination of characterizing the data as well as the data generation or data collection processes.

Challenges of Validating Inductive Learning

The performance of inductive learning methods can be tested by holding back some samples from the overall data set that has been collected and using those samples for validation. The presumption is that if the training set is used as the system requirements (the left-hand side of the V) an independent set of validation data can be used to ensure that the requirements have been met (forming the corresponding right-hand side of the V). Training data must not have accidental correlations unrelated to the desired behavior, or else the system will become "over-fitted." Similarly, the validation data must be independent and diverse from the training data in every way except the desired features, or else overfitting will not be detected during validation. It is unclear how to argue that a machine learning system has not been over-fitted as part of a safety argument.

A significant limitation of machine learning in practice is that if labelled data is used, each data point can be expensive. (Creating labels has to be done by someone or something. Unsupervised learning techniques are also possible, but require a clever mapping to solving a particular problem.) Moreover, if a problem with the training set (i.e., a requirements defect) or the learned rules (i.e., a design defect) is found and corrected, then more validation data has to be collected and used to validate the updated system. This is necessary because even a small change to the training data could produce a dramatically different learned rule set. Thus, complete revalidation would normally be required for any training set "bug fix," no matter how small.

Because of the complexity of requirements for an autonomous system, it seems likely that rare, edge cases will be where learning problems would be expected to occur. However, because of their rarity, collecting data depicting such unusual circumstances can be expensive and difficult to scale. (Simulation and synthetic data can help with this, but come with the risk of bias in simulated data, as well as overfitting to simulation artifacts.)

Another issue with validating machine learning is that, in general, humans cannot intuitively understand the results of the process. For example, the internal structure of a convolutional neural network [23] may not tell a human observer much intuitive about the decision rules that have been learned. While there might be some special cases, in general the problem of "legibility" [24, 25] of machine learning in terms of being able to explain in human terms how the system behaves is unsolved. This makes it difficult to predict how techniques other than expensive brute force testing can be applied for validation of machine learning systems. (Perhaps some organizations do have the resources to do extensive brute force testing. But even in this case the accuracy, validity, and representativeness of the training data must be demonstrated as part of any safety argument based on the correctness of a machine learning system.)

Because legibility for machine learning systems is generally poor, and because the danger of overfitting is real, there are failure modes in such a system that can significantly affect safety. Of particular concern are accidental correlations that are present in training set data but not noticed by human reviewers. For example, consider the method of detecting pedestrians in imagery using trained deformable-part models, which has been shown to be quite effective in real-world data sets [26]. If no (or few) images of pedestrians in wheelchairs were present in the training data set, it is likely that such a system would incorrectly correlate the label of "pedestrian" with "people who walk on two legs."

Solutions to Inductive Learning

Validating inductive learning is notoriously difficult due to the "black swan" problem [27], which is in general the susceptibility of a person (or system) to believe that common observations are true, and draw potentially incorrect conclusions due to an abundance of confirming data points. The story goes as follows. Before the late 1700s, all observed swans in Europe were white, and thus an observer using inductive logic would have concluded that all swans are white. However, this observer would experience a brittle failure of this belief when visiting Australia, where there are plenty of black swans. In other words, if there is a special case the system has not seen, it cannot learn that case. This is an essential limitation to inductive learning approaches that is not readily cured [28]. Moreover, with machine learning this problem is compounded by the lack of legibility, so it can be difficult or impossible for human reviewers to imagine what form a black swan-like bias in such a system might take.

Validating an inductive learning system seems to be an extremely challenging problem. Extensive testing might be used, but would require validating an assumption of random independent arrival rates of "black swan" data and testing on data sets sized accordingly. This might be feasible given enough resources, but there will always be new black swans, so a probabilistic assessment of huge numbers of operational scenarios and input values would have to be made to ensure an acceptably low level of system failures. (If resources were available to do this in a defensible way, this might suffice to form the right-hand-side of a V process.)

An alternative to validating inductive learning systems to high ASIL levels would be to pair a low-ASIL inductively-based algorithm that sends commands to an actuator with a high-ASIL deductively-based monitor. This would sidestep the majority of the validation problem for the actuation algorithm, since failures of the inductive algorithm controlling the actuator would be caught by a non-inductive monitor based on a concept such as a deductively-generated safety envelope. Thus, actuator algorithm failures would be an availability problem (the system safety shuts down, assuming an adequate failover capability) rather than a safety problem.

Mission Critical Operational Requirements

As a final technical area, we return to the previously discussed point that the computer is ultimately in control of the vehicle rather than the person being in control. That means that at least some portion of the vehicle has to be fail-operational rather than fail-stop.

Challenges of Fail-Operational System Design

Fail operational system design has been done successfully in aerospace and other contexts for decades, but is still difficult for several reasons. The first reason is the obvious one that redundancy has to be provided so that when one component fails another one can take over. Achieving this requires at least two independent, redundant subsystems for fail-stop behavior.

Achieving a fail-operation system in turn requires at least three redundant fail-arbitrary components so that it can be determined which of the three failed in the event that it issues incorrect outputs rather than failing silent at the component level [29]. For systems that have to tolerate arbitrarily bad faults, a Byzantine fault tolerant system with four redundant components might be required [30], depending on the relevant fault model.

The structure of the redundancy varies depending on the design approach, and might include configurations such as a triplex redundant system with a voter (in which case the voter must be ensured not to be a single point of failure), or a dual two-of-two system that uses four computers in fail-silent pairs [29]. Beyond the obvious expense such approaches introduce, there is also an issue of testing to make sure that failure detection and recovery works, assuring independence of failure, and ensuring that all redundant components are fault-free at the start of a driving mission. It seems unlikely that redundancy can be avoided, but it may be possible to reduce the complexity and expense of providing sufficient redundancy to ensure safety.

Failover Missions

In typical fail-operational system such as aircraft, all the redundant components are essentially identical and capable of performing an extended mission. For example, commercial aircraft are commonly configured with two jet engines, and each jet engine has at least a dual-redundant computer control. If the pair of computers on one engine shuts down due to a fault detected via continual crosschecking, there is a second independent engine to keep the aircraft flying. Even so, the requirements on engine dependability are very stringent, because aircraft might potentially have to fly several hours after a first engine failure to reach the nearest airport without having the second engine fail. This puts significant reliability requirements on each engine, and therefore increased component costs.

While cars are notoriously cost-sensitive, they do have an advantage in that failover missions can be short (e.g., pull over to the side of the road, or if necessary come to a stop in a travel lane), with failover mission durations measured in seconds rather than hours. Additionally, a failover mission to stop the vehicle might be able to operate with significantly less functionality than fully autonomous operation. This can simplify requirements complexity, computational redundancy, sensor requirements, and validation requirements. (As a simple example, a failover mission control system might not support lane changes, greatly simplifying sensor requirements and control algorithms. More sophisticated approaches that are still simpler than full autonomy might be possible.) Therefore, designing an autonomous vehicle with a fail-stop primary controller and a simpler fail-operational failover controller might be attractive both in terms of hardware cost and in terms of design/validation cost.

It might also be that a safety argument can be created not based on the full autonomy system being perfect, but rather on the full autonomy system having a detector that realizes when it is malfunctioning or has encountered a gap in its requirements. This would make the fault detector itself high-ASIL, but might permit normal autonomy functions to be low-ASIL. Such an approach would map well onto a monitor/actuator architecture for the primary autonomy system. The failover autonomy would also have to be designed in a safe manner, with an appropriate architectural approach depending on its complexity and calculated reliability requirements. It might even be possible to use a single-channel failover system if the probably of failure during a short failover mission lasting only seconds is sufficiently low.

Non-Technical Factors

Some challenges in deploying autonomy are non-technical, such as the frequently mentioned liability problem (who pays when there is a mishap?) and how laws generally treat the ownership, operation, maintenance, and other aspects such vehicles.

A deep dive into this topic is beyond the scope of this paper. However, resolutions to non-technical challenges will very likely have an impact on technical solutions. For example, there may be forensic requirements imposed on autonomy systems for accident reconstruction data. Careful analysis of the provenance of such data will need to be performed to ensure that the data is used properly. As a simple example, if a radar has a hypothetical detection probability of 95%, its output might still be recorded in the system in terms of whether an obstacle was or was not detected, superficially implying detection certainty. It is important to ensure that forensic analysis takes into account that just because the radar didn't detect a pedestrian does not mean the pedestrian was not there (e.g., a 95% detection probably implies that 1 out of 20 pedestrians will not actually be detected).

It seems likely that with the inherent complexity of an autonomous vehicle and the clear inability to demonstrate anything close to perfection via testing, it will be important for developers to create a safety assurance argument in the form of an assurance case (e.g., according to [31]). Such an assurance argument will be necessary to defend and explain the integrity of their system and be able to credibly explain the system's responses to events surrounding the inevitable mishaps that will occur. A particular point that should be addressed is ensuring the integrity of evidence to establish whether a mishap was reasonably unavoidable due to its circumstances. Other important points will be whether or not a mishap was arguably caused by a defect in system requirements (e.g., a gap in training data), a reasonably foreseeable and avoidable design defect, an implementation defect, or other cause attributable to the vehicle manufacturer.

Fault Injection

As is apparent from the preceding discussion, traditional functional testing will have trouble exercising a complete system, and especially will find it difficult to exercise combinations of exceptions occurring during unusual operational conditions. While testers can define some off-nominal test cases, scalability of that testing is questionable due to the combinatorial explosion of exceptions, operational scenarios, and other factors involved. Additionally, it has been shown that even very good designers often have blind spots and miss exceptional situations in comparatively simple software systems [32].

Fault injection and robustness testing are relatively mature technologies for assessing the performance of a system under exceptional conditions [33], and can help avoid designer and tester blind spots when testing exceptional condition responses. Traditional fault injection involves inserting bit flips into memory and communication networks. More recent techniques have increased the level of abstraction to include data-type-based fault dictionaries [32], and ensuring fault representativeness [33]. Such techniques have already been used successfully to find and characterize defects on autonomous vehicles [35].

A promising approach to helping validate autonomy features is to perform fault injection at the level of abstraction of the component, as part of a strategy of attempting to falsify claims of safety [36]. This involves not only simulating objects for primary sensor inputs, but also inserting exceptional conditions to test the robustness of the system (e.g., inserting invalid data into maps). The point of doing such fault injection is not to validate functionality, but rather to probe for weak spots that might be activated via unforeseen circumstances. Such fault injection can be performed across the range of layers in the ISO 26262 V process [37].

Conclusions

The challenges of developing safe autonomous vehicles according to the V process are significant. However, ensuring that vehicles are safe nonetheless requires following the ISO 26262 V process, or demonstrating that a set of process and technology practices equally rigorous has been applied. Assuming that the V process is applied, there are three general approaches that seem promising.

Phased Deployment

It appears impractical to develop and deploy an autonomous vehicle that will handle every possible combination of scenarios in an unrestricted real-world environment, including exceptional situations, all at once. Rather, as is common in automotive systems, a phased deployment approach building on current developer practice seems likely to be a reasonable approach.

Tying phased deployment to the V process can be done by identifying well-specified operational concepts to limit the scope of operations and therefore the necessary scope of requirements. This would include limitations in environment, system health, and operational constraints that must be satisfied to enable autonomous operation. Validating that such operational constraints are enforced will be an essential part of ensuring safety, and will have to show up in the V process as a set of operational requirements, validation, and potentially run-time enforcement mechanisms. For example, run-time monitoring might be required to monitor not only whether system state is in a permissible autonomy regime, but also that assumptions made about the operational scenario in the safety argument are actually being satisfied, and whether the system is actually in the operational scenario it thinks it is in.

An aspect of restricted operational concepts that will require particular attention is ensuring that safety is maintained when an operational scenario suddenly becomes invalidated, due to for example an unexpected weather event or an infrastructure failure. Such exceptional transitions out of an acceptable operational concept regime will require that system recovery or a failover mission be executed successfully even when there is a system excursion outside the assumptions of permissible autonomy operational scenarios.

It is unclear whether a phased deployment approach will provide a path all the way to complete autonomy. But at least such an approach provides a way to make progress and gain some benefits of autonomy while gaining greater understanding of the difficult edge cases and unanticipated scenarios that will arise as systems see more exposure to real world conditions.

Monitor/Actuator Architecture

A common approach that might help mitigate many of the challenges of autonomous vehicle safety is the use of a monitor/actuator architecture. As discussed, this architectural style can help with requirements complexity (only the monitor needs to be essentially perfect), and deployment of inductive algorithms (by limiting use of induction to the actuator, and using a deductively-based monitor).

Additionally, the use of a failover mission strategy can allow a primary autonomy system monitor to detect a primary system failure without having to ensure fail-operational behavior. A simpler, high-integrity failover autonomy system can bring the vehicle to a safe state. Such a system might have a failover mission short enough that minimal redundancy for failover operation is required, so long as it can be assured that the system is fault-free when it is time to start a failover mission.

Fault Injection

Testing alone is infeasible to ensure ultra-dependable systems. Autonomous vehicles only make this problem harder by automating responses to highly complex environmental situations, and introducing technology such as machine learning that is difficult and expensive to test. Moreover, because much of the autonomy capability must have a high ASIL due to the lack of human driving oversight, it seems difficult to do enough ordinary system testing to gain even a reasonable level of assurance.

Fault injection can play a useful role as part of a validation strategy that also includes traditional testing and non-test-based validation. This is especially true if fault injection is applied at multiple levels of abstraction rather than just at the level of stuck-at electrical connectors.

Future Work

This paper discusses ways to fit autonomous vehicle safety assurance within an ISO 26262-based V framework. However, it is expected that using architectural patterns such as the monitor/actuator approach and the practical limits of validation possible via fault injection will place constraints on operational performance. In other words, the functionality of autonomous vehicles might need to be limited to fit the constraints of feasible validation techniques. Relaxing those constraints will require advances in areas such as characterizing the coverage of machine learning training data compared to the expected operational environment, gaining confidence in safety requirements with regard to exceptional driving conditions, and being able to validate the independence of failures in redundant inductive-based systems.

Contact Information

Dr. Philip Koopman is an Associate Professor of Electrical and Computer Engineering at Carnegie Mellon University, where he specializes in software safety and dependable system design. He also has affiliations with the National Robotics Engineering Center (NREC) and the Institute for Software Research.
E-mail: koopman@cmu.edu.

Michael Wagner is the CEO and co-founder of Edge Case Research, LLC, which specializes in software robustness testing and high quality software for autonomous vehicles, robots, and embedded systems. He is also has an affiliation with the National Robotics Engineering Center.
E-mail: mwagner@edge-case-research.com

Definitions/Abbreviations

ADAS - Advanced Driver Assistance System
ASIL - Automotive Safety Integrity Level
HOV - High Occupancy Vehicle
LIDAR - Light Detection and Ranging
V model - A software development model that includes requirements and design on the left side of a "V" with verification and validation on the right side of the "V"

References

1. Transportation Research Board, *National Automated Highway System Research Program: A Review*, TRB Special Report 253 (Washington, DC: National Academy Press, 1998).

2. NHTSA, "Preliminary Statement of Policy Concerning Automated Vehicles," May 2013, accessed October 2015, http://www.nhtsa.gov/staticfiles/rulemaking/pdf/Automated_Vehicles_Policy.pdf.

3. US Department of Transportation, FAA, Advisory Circular, "System Design and Analysis," AC 25.1309-1A, June 21, 1988.

4. Butler and Finelli, "The Infeasibility of Experimental Quantification of Life-Critical Software Reliability," *IEEE Trans. SW Engr.* 19, no. 1 (January 1993): 3-12.

5. Motor Industry Software Reliability Association, "Development Guidelines for Vehicle Based Software," November 1994.

6. Motor Industry Software Reliability Association, "Report 6: Verification and Validation," February 1995.

7. "Road Vehicles—Functional Safety—Part 2: Management of Functional Safety," ISO 26262-2:2011, November 15, 2011.

8. "Road Vehicles—Functional Safety—Part 3: Concept Phase," ISO 26262-3:2011, November 15, 2011.

9. "Road Vehicles—Functional Safety—Part 9: Automotive Safety Integrity Level (ASIL)-Oriented and Safety-Oriented Analyses," ISO 26262-9:2011, November 15, 2011.

10. Randell, B., "System Structure for Software Fault Tolerance," IEEE Trans. SW Engineering SE-1:2, June 1975, 1-18.

11. Lions, J., "ARIANE 5 Flight 501 Failure," Report by the Inquiry Board, Paris, July 1996.

12. Yeh, Y.C., "Design Considerations in Boeing 777 Fly-by-Wire Computers," HASE, 1998.

13. Wereschagin, M., "Pittsburgh Left' Seen by Many as a Local Right," *TribLive News*, June 14, 2006, http://triblive.com/x/pittsburghtrib/sports/s_457936.html.

14. Bayouth, M. and Koopman, P., "Functional Evolution of an Automated Highway System for Incremental Deployment," Transportation Research Record, #1651, Paper #981060, 80-88.

15. Shladover, S. et al., "Development and Performance Evaluation of AVCSS Deployment Sequences to Advance from Today's Driving Environment to Full Automation," California PATH Research Report UCB-ITS-PRR_2001-18, August 2001.

16. Black, J. and Koopman, P., "System Safety as an Emergent Property in Composite Systems," DSN, 2009, 369-378.

17. Kane, Chowdhury, Datta, and Koopman, "A Case Study on Runtime Monitoring of an Autonomous Research Vehicle (ARV) System," RV, 2015.

18. Geraerts, R. and Overmars, M.H., "A Comparative Study of Probabilistic Roadmap Planners," *Proc. Workshop on the Algorithmic Foundations of Robotics (WAFR'02)*, 2002, 43-57.

19. Martin, C. and Moravec, H., "Robot Evidence Grids," tech. report CMU-RI-TR-96-06, Robotics Institute, Carnegie Mellon University, March, 1996.

20. Dollár, P. et al., "Pedestrian Detection: An Evaluation of the State of the Art," *IEEE Trans. on Pattern Analysis and Machine Intelligence* 34, no. 4 (2012).

21. Silver, D., Bagnell, J, and Stentz, A., "Active Learning from Demonstration for Robust Autonomous Navigation," *IEEE Conference on Robotics and Automation*, May 2012.

22. Dima, C., "Active Learning for Outdoor Perception," doctoral dissertation, tech. report CMU-RI-TR-06-28, Robotics Institute, Carnegie Mellon University, May 2006.

23. Krizhevsky, A., Sutskever, I., and Hinton, G.E., "ImageNet Classification with Deep Convolutional Neural Networks," *NIPS*, 2012, 1106-1114.

24. Dosovitskiy, A. and Brox, T., "Inverting Convolutional Networks with Convolutional Networks," *CoRR* abs/1506.02753 (2015).

25. Zeiler, M.D. and Fergus, R., "Visualizing and Understanding Convolutional Networks," *ECCV*, 2014.

26. Felzenszwalb, P., Girshick, P., McAllester, D., and Ramanan, D., "Object Detection with Discriminatively Trained Part Based Models," *IEEE Transactions on Pattern Analysis and Machine Intelligence* 32, no. 9 (September 2010).

27. Taleb, N., *The Black Swan: The Impact of the Highly Improbable, Random House*, 2007.

28. Hume, D., *An Enquiry Concerning Human Understanding* (P.F. Collier & Son, 1910 [1748]), ISBN 0-19-825060-6.

29. Hammet, "Design by Extrapolation: An Evaluation of Fault-Tolerant Avionics," *IEEE Aerospace and Electronic Systems* 17, no. 4 (2002): 17-25.

30. Lamport, L., Shostak, R., and Pease, M., "The Byzantine Generals Problem," *Trans. Prog. Lang. Sys.* 4, no. 3 (July 1982): 382-401, ACM.

31. "Systems and Software Engineering—Systems and Software Assurance—Part 2: Assurance Case," ISO/IEC 15026:2011.

32. Koopman, P., DeVale, K., and DeVale, J., Interface Robustness Testing: Experiences and Lessons Learned from the Ballista Project, Kanoun, K. and Spainhower, L. Eds., *Dependability Benchmarking for Computer Systems* (IEEE Press, 2008), 201-226.

33. Kanoun, K. and Spainhower, L. Eds., *Dependability Benchmarking for Computer Systems* (IEEE Press, 2008), 201-226.

34. Natella, R., Cotroneo, D., Duraes, J., and Madeira, H., "On Fault Representativeness of Software Fault Injection," *IEEE Trans. SW Engineering*, 39, no. 1 (January 2013): 80-96.

35. Vernaza, P., Guttendorf, D., Wagner, M., and Koopman, P., "Learning Product Set Models of Fault Triggers in High-Dimensional Software Interfaces," *IROS*, 2015.

36. Wagner, M. and Koopman, P., A Philosophy for Developing Trust in Self-Driving Cars, Meyer, G. and Beiker, S. Eds., *Road Vehicle Automation 2, Lecture Notes in Mobility* (Springer, 2014), 163-170.

37. Pintard, L., Fabre, J.-C., Kanoun, K., Roy, M. et al., "Fault Injection and Automotive Development Process," *Embedded Real-Time Software And Systems* (*ERTS2*), Fev. 2014, Toulouse, France.

RV-ECU: Maximum Assurance In-Vehicle Safety Monitoring

Philip Daian, Bhargava Manja, and Grigore Rosu
Runtime Verification Inc.

Shinichi Shiraishi
Toyota Info Technology Center USA

Akihito Iwai
DENSO International America Inc.

The Runtime Verification ECU (RV-ECU) is a new development platform for checking and enforcing the safety of automotive bus communications and software systems. RV-ECU uses runtime verification, a formal analysis subfield geared at validating and verifying systems as they run, to ensure that all manufacturer and third-party safety specifications are complied with during the operation of the vehicle. By compiling formal safety properties into code using a certifying compiler, the RV-ECU executes only provably correct code that checks for safety violations as the system runs. RV-ECU can also recover from violations of these properties, either by itself in simple cases or together with safe message-sending libraries implementable on third-party control units on the bus. RV-ECU can be updated with new specifications after a vehicle is released, enhancing the safety of vehicles that have already been sold and deployed.

Currently a prototype, RV-ECU is meant to eventually be deployed as global and local ECU safety monitors, ultimately responsible for the safety of the entire vehicle system. We describe its overall architecture and implementation, and demonstrate monitoring of safety specifications on the CAN bus. We use past automotive recalls as case studies to demonstrate the potential of updating the RV-ECU as a cost effective and practical alternative to software recalls, while requiring the development of rigorous, formal safety specifications easily sharable across manufacturers, OEMs, regulatory agencies and even car owners.

CITATION: Daian, P., Shiraishi, S., Iwai, A., Manja, B. et al., "RV-ECU: Maximum Assurance In-Vehicle Safety Monitoring," SAE Technical Paper 2016-01-0126, 2016, doi:10.4271/2016-01-0126.

Introduction

Modern automobiles are highly computerized, with 70 to 100 complex and interconnected electronic control units responsible for the operation of automotive systems, and roughly 35 to 40 percent of the development cost of modern automobiles going towards software. In the next 10 years this number is expected to jump to between 50 and 80 percent, and even higher for hybrid vehicles. This will only be more true with the advent of autonomous vehicles [1, 2].

It is not surprising, then, that the automotive industry suffers from nearly every possible software fault and resulting error. Many related stories have recently been featured on the news, including cases where cars are hacked and remotely controlled, including brakes and the engine, completely ignoring driver input. In some cases prior physical access to the car was needed, in others the car was not even touched. Massive automobile recalls in the past few years have been due to software bugs, costing billions [3, 4, 5, 6, 7, 8, 9]. Moreover, almost 80 percent of car innovations currently come from computer software, which has therefore become the major contributor of value in cars [1]. As software becomes more and more integral to the function and economics of vehicles, the safety and security of car software has taken center stage.

Limitations of Current Approaches

Traditional software development quality processes rely on static analysis tools and techniques to improve the quality, security and reliability of their code. Static analysis tools analyze software code against a set of known rules and heuristics and notify the operator of warnings and violations. Nearly all companies developing a reasonably large code base use code quality tools. The reader interested in how static analysis tools perform on automotive-related software is referred to [10]. Even with all of the resources spent on these tools, software is still full of bugs and reliability weaknesses. This may be fine when the software is running on something as simple as a cell phone, or a laptop computer that your child uses for homework, but this is unacceptable at best, and dangerous at worst, when the software runs in an automobile.

Model checking [11] is a complementary approach that has found some use in the automotive industry. While rigorous and thorough, this approach suffers from serious drawbacks that make its use impractical. Besides the infamous "state explosion" problem, the most significant drawback of model checking is the issue of model faithfulness. Models being used must be correct with regards to the system being inspected and the environment it operates in. With the complexity of modern software and hardware systems, and the (often) specific nature of the models involved, great care must be taking in validating the model itself as well as the system with regards to the model. This is an extremely error prone and time intensive process. A previous comparison of model checking to static analysis by a team investigating model checking tools found that issues in the model itself caused model checking to miss five errors caught by static analysis, concluding that "the main source of false negatives is not incomplete models, but the need to create a model at all. This cost must be paid for each new checked system and, given finite resources, it can preclude checking new code." [12]. Besides the model, the tool itself must also be trusted to properly verify the properties over the model, requiring either a highly-audited open source tool or another source of high confidence in the tool itself.

The portability of these models and specifications is also dubious: any changes in the underlying system require a correct change in the model, a non-trivial process that must be repeated often for complex systems [11]. Equivalent specifications can thus have different meanings based on the models being used.

While this does not matter if the model is specific to some standard, such as a programming language [13], with many tools and applications of model checking this is not the case [14, 15]. So, while expressive, models can be complex and non-portable. Overall, while model checking has the potential for detecting deep and subtle errors, the requirement for a model introduces many restrictions and complexities that make the tools difficult to manage and integrate effectively into most engineering teams, restricting their use to teams with high levels of formal expertise and critical applications requiring the maximum possible assurance, thus preventing widespread adoption by the automotive industry as a whole.

Enabling Safety Standardization

Another hurdle on the path to greater automotive safety is the lack of standardized automotive safety specifications. Because many specifications are informally expressed and never formalized, communication between Tier 1 suppliers and their OEM partners is often incomplete with regards to safety, producing components that may behave unpredictably in the system as a whole. Moreover, formalizations that exist tend to be difficult or impossible to port between Tier 1 suppliers. One clear industry need stemming from verification-based development methodologies is the need for portable formal safety specifications. Specifications should be expressed in lightweight formalisms that are easy to understand and communicate, and should stay separate from the particular verification approach that is employed for their checking.

Lastly, we observe that in currently developed automotive systems, both the safety and the functionality of the system and its components are considered and implemented together, as part of the same development process. Because safety and functionality are necessarily related to each other, this appears to be logical. However, this intermixing of safety checks in components that are primarily functional represents a violation of the maximum possible separation of concerns in an ideal system architecture, in which safety would be considered and implemented separately from the desired functionality, allowing for a clean separation that promotes both safety testing and rigorous reasoning about safety properties.

As an alternative to static verification methods, run-time verification makes possible easy standardization of rigorous formal safety specifications and clean separation between functionality and safety components of systems.

Runtime Verification

Runtime verification is a system analysis and approach that extracts information from the running system and uses it to assess satisfaction or violation of specified properties and constraints [16]. Properties are expressed formally, as finite state machines, regular expressions, linear temporal logic formulas, etc. These formal requirements are used to synthesize monitors, and the existing code base is automatically instrumented with these monitors. Runtime verification can be used for many purposes, including policy monitoring, debugging, testing, verification, validation, profiling, behavior modification (e.g., recovery, logging), among others. In the development cycle, runtime verification can be used as a bug finding tool, a testing approach, a development methodology focusing on the creation of formally rigorous specifications, while in a production system it can be used as a component responsible for enforcing a set of safety requirements on a system to preserve its global safety during operation.

Ideally, developers and software quality managers would like to validate a system prior to its operation and release, in the coding and testing phases of the software engineering lifecycle. This would allow maximum assurance in the performance of the deployed system before release, increasing software dependability. However, as previously discussed, static validation methods such as model-checking [17] suffer from limits preventing their use in real large-scale applications. For instance, those techniques are often bound to the design stage of a system and hence they are not shaped to face-off specification evolution. Even when static analysis techniques do scale, they are limited by the properties they can check, and may not be able to check interesting behavioral properties. Thus, the verification of some properties, and elimination of some faults, have to be complemented using methods relying on analysis of system executions.

Figure 1 shows an example of an automotive safety specification being compiled to code that enforces it at runtime using the technology underlying RV-ECU. The specification is called "safe door lock", and is first stated in plain English informally, as safety requirements are currently expressed. This is translated to a formal requirement manually by domain experts, as shown in the orange box using linear temporal logic: it is always the case that a valid door open event implies that there has not been a door lock since the last unlock; a recovery action is attached that closes the door when a violation is detected (violation handler).

Previous efforts in the runtime verification field have focused on the development of formalisms appropriate for specifying expressive properties while synthesizing efficient monitors [18, 19, 20, 21, 22, 23], steering program and system executions to obtain desirable behaviors [24], combining runtime verification with other approaches including static analysis [25], minimizing runtime overhead to make monitoring deployed systems practical [22, 26], and integrating run-time verification with existing projects automatically through both binary and source instrumentation, often leveraging aspect-oriented programming [27, 28].

Because runtime verification is a relatively new field, the number of practical and commercial applications of the technology is less substantial than that of static analysis

FIGURE 1 Example automotive specification being compiled into code that enforces it at runtime through RV-ECU.

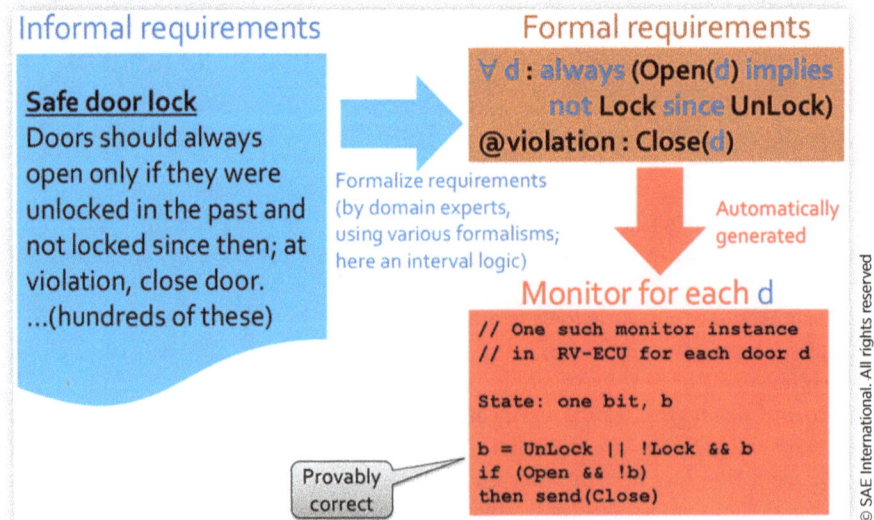

tools, model checkers, or deductive program verifiers. There have, however, been some practical applications of the theory of runtime enforcement for program safety and security [29, 30, 31], or to enforce access control policies at system runtime [32, 33]. Runtime verification has also been applied to mobile applications to provide fine-grained permissions controls and enforce device security policies [34].

RV-ECU: A Vehicle Safety Architecture

Because the automotive industry develops some of the most widely deployed safety critical software of any industry, it represents an ideal context where the benefits of runtime verification can make a significant difference.

Towards this goal we introduce RV-ECU, a development platform, also referred to as a "workbench" or a "system" in the paper, for checking and enforcing the safety of automotive bus communications. For brevity, whenever the context is non-ambiguous we take the freedom to use the same name "RV-ECU" for any of its components or even for other components that make use of code produced using RV-ECU.

At its core, RV-ECU consists of a compiler from formally defined safety specifications to monitoring code running on embedded control units. The safety specifications can be designed in any known mathematical formalism, with RV-ECU providing a plugin-based system to enable the development of custom formalisms for the specific automotive needs. Currently, some supported specifications languages include finite state machines, regular expressions, linear temporal logic, and context-free grammars.

To provide a clearer picture of what RV-ECU is and what it can do, we will explain its use, from end to end and step by step:

1. Trained personnel use the formalism of their choice to specify a safety property. Take, for example, the property that the windshields being on the fastest setting implies the headlights are on the brightest setting. The specification includes "recovery actions" to take if the property is violated, to return to a safe state

2. The RV-ECU compiler is invoked on the formal definition, creating monitors and proof objects that certify the monitors are correct with regards to the specification

3. The original code base is instrumented, either automatically or manually, with calls to monitors are relevant call sites. Thus, the safety checking/recovery functionality is cleanly orthogonal to other functionality considerations.

Figure 2 shows an overview of the RV-ECU methodology, which takes formal specifications as input and from them automatically outputs code checking these specifications as well as a proof object certifying the correctness of this code over the mathematical semantics of the specification formalism and of the underlying programming language. Thus, the code output by the RV-ECU compiler provides correctness proof certificates of the monitoring code as well as of the recovering code which is executed when the original specification is violated. These certificates can be checked in third party theorem proving software, providing the maximum known assurance guarantees that the code running on-device implements the given safety specifications and their recovery handlers.

The benefits of the RV-ECU approach are numerous. RV-ECU's compatibility with many formalisms and its function as a compiler to monitors completely decouples considerations of functionality from those of safety. Automotive software engineers are free to focus their efforts on code that enhances the functionality of software systems

FIGURE 2 RV-ECU system, applying automatic certifying compilation of safety specifications.

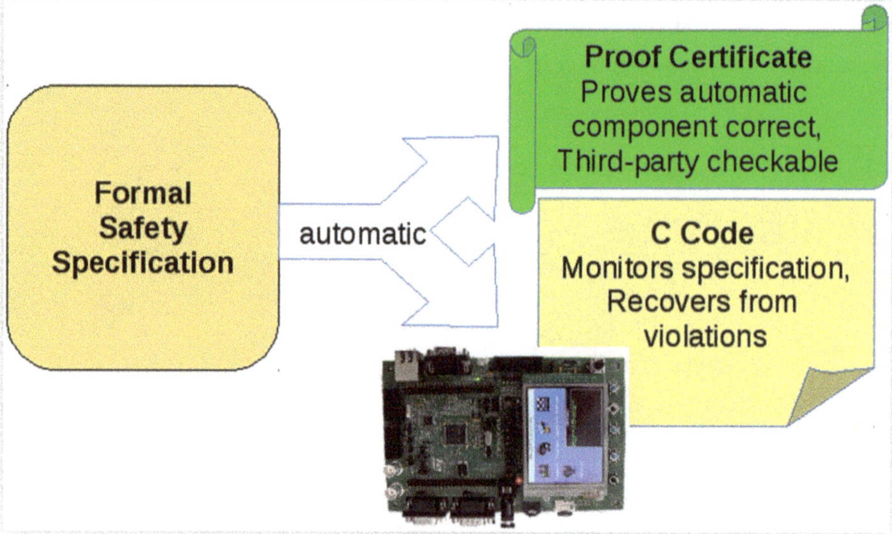

aboard the automobile, while safety engineers can focus on formalizing and testing safety properties. This decoupling allows for the development of modular and reusable safety formalisms that can easily be shared between automotive suppliers and OEMs. This can be revolutionary, as it ensures compatibility in safety specifications between OEMs and Tier 1 suppliers. It even makes possible a standardized database of formal safety properties maintained and updated by state regulatory bodies.

Global and Local Monitoring

Figure 3 shows the RV-ECU system running on a vehicle. It is important to note that RV-ECU can be applied in two places: the generated monitoring code can either run on a separate control unit to monitor the global traffic on the bus, or be integrated within an existing control unit (e.g., the power steering ECU) to prevent it from taking unsafe actions. We therefore distinguish two categories of monitors, with "global" monitors observing the bus on a dedicated ECU and "local" monitors observing the bus from the perspective of an existing ECU responsible for some functionality.

These global and local monitors can then further communicate to increase their effectiveness. When used together, the global monitors can track the state of the entire vehicle system, with local monitors tracking only the state important to a particular controller. By communicating over the CAN bus, they are able to share messages and commands, and the global monitor is able to instruct the local monitors to block or modify messages they may otherwise allow.

For simple testing and safety specifications involving one component, local monitoring can be used. With complex or resource-intensive properties involving multiple components, global monitoring can be used. A combination of these approaches can be applied both in the testing cycle and the production automobile, spanning the extremes between global monitoring of the entire system only with untrusted code running on individual components and local monitoring of specific components only with no global specifications or dedicated hardware. This flexibility allows OEMs and Tier 1 suppliers to choose how and where they apply the runtime verification technology,

FIGURE 3 RV-ECU running both globally and locally, checking and enforcing vehicle system-wide safety.

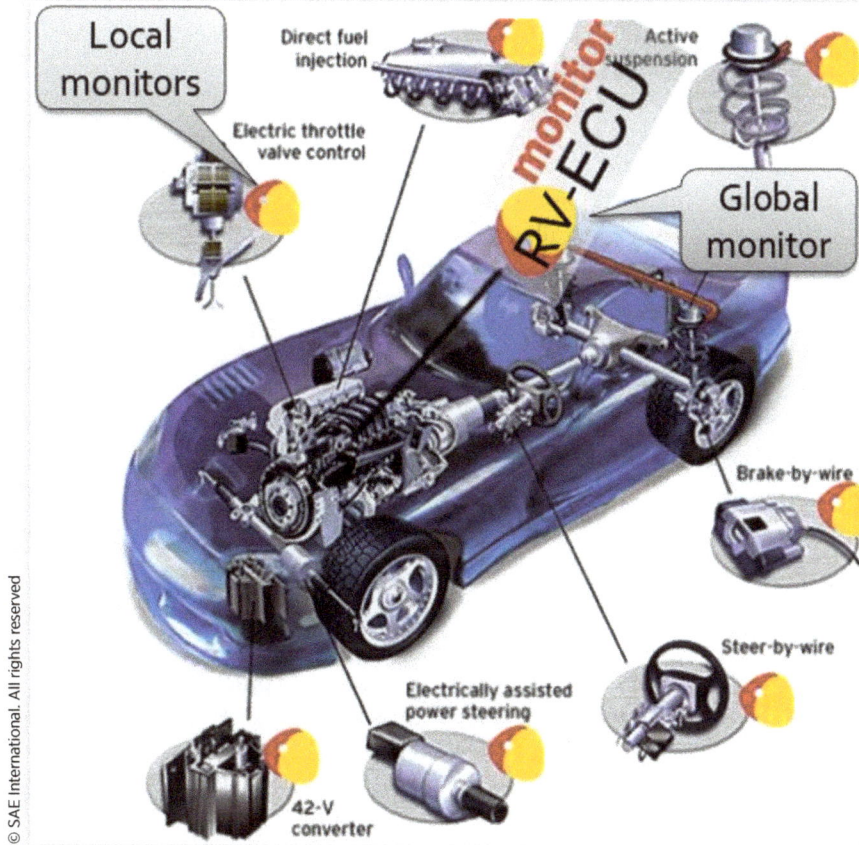

allowing for incremental rollouts of local monitors at first followed by the eventual implementation of a global monitor, or vice versa.

Figure 4 shows the ideal RV-ECU deployment, with all ECUs on the bus containing local monitors and a global monitor attached to the full system. In this example, no communication can flow between untrusted, manually-written code implementing functionality (highlighted in yellow) and the vehicle bus without approval from high-assurance, provably correct, automatically generated code implementing the safety specifications of the vehicle.

The use of RV-ECU therefore protects the overall safety of the system from both malfunctioning controllers and malicious accesses (hackers), maintaining a set of safety invariants specified rigorously during the development of the vehicle. Moreover, the safety monitoring code generated by RV-ECU uses state-of-the-art techniques and

FIGURE 4 RV-ECU protecting the CAN bus from unsafe messages.

algorithms developed by the runtime verification community specifically aimed at minimizing runtime overhead.

Certifiable Correctness

As previously mentioned, the code generated by the RV-ECU system from safety specifications additionally carries proof certificates. Proof certificates are mathematical objects expressed as objects in the Coq automated theorem proving assistant [35], a proof assistant that has been widely successfully applied to detect security flaws in popular software [36], prove mathematical theorems [37], and create and prove the most complete currently certified C compiler [38].

The proof objects we provide mathematically prove that the code we generate correctly implements the specification inputs provided, with regards to the mathematical formal semantics of the programming language itself. These proofs are machine checkable by third party theorem proving tools including but not limited to Coq, providing multiple independent sources of assurance that the generated code is rigorously correct. Such proof objects can also be used in the context of certifying vehicle safety, with their formal rigor providing the maximum known standards for software development in the context of rating development assurance in standards like ISO 26262.

To demonstrate such proof certificates, we have proved the simple finite state monitor shown in Figure 5 using our in-house verification technology, the K Framework. The figure shows the transition function of the finite state monitor, in this case a simplified transition function implemented in KernelC. It also shows a proof specification for the program, showing that we would like to prove that the main function rewrites to a void value. This void value in our semantics of KernelC indicates that the program execution terminated with no error.

In this example we prove that the program, when instrumented with the monitoring code, will never reach a state we define as an error state. Our proof thus requires both the monitoring code and the program being instrumented to complete, and proves the correctness of our recovery action with regards to our property as well.

We reached several challenges in defining the target for such a proof of correctness. Our initial operating assumption was that a monitor is correct when, for all possible event traces, the monitoring code will end in an internal state consistent with the property it claims to monitor. This definition does not however take into account the steering aspect

FIGURE 5 FSM transition for monitor and formal specification verying program.

```
void step(int *state, int event) {          require "../../kernelc.k"
    if(*state == 0) {
        if(event == 0) {                     module MINIMUM-SPEC
            *state = 1;                         imports KERNELC
        } else if(event == 1) {
            *state = 2;                       rule
        }                                       <fun>... FUN:Map ...</fun>
    } else if(*state == 1) {                    <k>
        if(event == 0) {                          main(.Expressions)
            *state = 3;                         =>
        } else if(event == 1) {                   tv(void, undef)
            *state = 0;                         </k>
        }
    }                                         endmodule
}
```

of runtime monitors: because runtime monitors are interactive in the program, they have the ability to modify the program's path as they run and execute recovery actions. Some event traces that may be captured in the property may then be unreachable by the monitor, which actively steers the program away from such traces through recovery actions.

Another key point that is not captured by the above definition of monitor correctness is instrumentation. If it is possible that a monitor may miss events, any guarantees provided by the above definition are entirely meaningless. If a monitor process events that did not occur, the same is true. Our final correctness notion somehow thus should include the notion of instrumentation.

Naturally, we need to draw a line somewhere to create a trusted assurance base usable in what we seek to prove. We believe the compiler boundary is the ideal place for that separation in our work: we assume that the compiler will not introduce any behavior into the system inconsistent with the semantics of the programming language it takes as input, and we assume a lack of hardware faults. Increasing the assurance of both areas is a separate research task with ongoing academic research being pursued in certified compilation [39] [40] and trusted hardware systems [41] [42] [43].

Because of the need to include instrumentation in our correctness guarantees, we must therefore naturally consider the program being executed (which is the instrumentation site for the monitor). We believe the correct definition of a correct (Monitor, Program) pair is then that the Program, when instrumented with the Monitor, will never reach any state violating the specified safety property. Thus, we must prove that the safety property itself holds over the program, which is equivalent to proving that the safety property holds in the unmonitored program. This is exactly the proof we created through our K Framework verification technology, using the inputs shown in Figure 5 as well as the C code of the same program being executed.

We are continuing to explore alternate correctness definitions and guarantees, and are developing the infrastructure to provide Coq-verifiable certificates for all potential guarantees.

RV-ECU Compared: Other RV Efforts

There is a fair amount of supporting work in applications of runtime verification to critical systems, embedded systems, and automotive systems. The most similar to our work is by Kane, wherein runtime CAN bus monitors for a range of properties are implemented [44]. However, his work is not viable for industrial use. His chosen development board, an STFF4-Discovery microcontroller, does not include a CAN transceiver, so he added a breadboard for CAN read/write functionality. By contrast, RV-ECU is an integrated hardware/software system, with the hardware ECU is capable of being used in a vehicle without modification, as we have shown in our demo.

Other work in runtime monitoring for ultra critical systems and hard real time monitoring also falls short in its applicability to the automotive sector. One representative example is the Copilot system from Galois, Inc. [45] [46]. Though it, too, is a compiler from formalisms to embedded C monitors, it requires a completely custom formalism to specify properties, and moreover requires expertise in Haskell, a niche language, and the use of the custom Copilot embedded domain specific language. In contrast, RV-ECU's modular plugin system allows specification in arbitrary formalisms and knowledge only of C, the language of choice in automotive software.

Many other runtime systems for automotive monitoring require specialized hardware or hardware modifications to ECUs [47] [48]. In contrast, RV-ECU can function completely in software, as the actual monitoring ECU is a completely optional addition to the CAN network. Of all systems we have seen, RV-ECU is the most flexible, adaptable, and general system for runtime monitoring of automobiles. It also currently has the most rigorous infrastructure for proving monitors correct, giving some additional assurance that the code on-device behaves as intended. We intend to continue extending these features as they become relevant to our partners and customers.

Recalls and RV-ECU, a Case Study

One of the key problems in the automotive industry we believe will be helped by the RV-ECU technology is a reduction in the required number of software recalls, as well as a quicker and less costly response when recalls must be performed. To demonstrate this application of our technology, we consider previous software-caused recalls in the automotive industry.

We do not have to look far to find good examples. Just a few months ago two security researchers unveiled an exploit that gave them full, remote access to the CAN bus of the Chrysler Jeep Cherokee [49]. The two researchers found an unauthenticated open port on the car's Uconnect cellular network interface, and used this foothold, as well as the fact that firmware binaries were unsigned, to update the car's networking hardware over the air with a backdoored firmware that gave them the ability to sniff CAN messages.

An unauthenticated SPI line between their back-doored chip and a CAN controller allowed them to write arbitrary messages over the CAN bus. Their control over the car was near total - they demonstrated complete wireless control over braking, the sound system, the driver display, door locks, AC, windshield wipers, steering (in reverse), and transmission [50]. They publicized their research, after disclosing the issue to regulators the car companies involved, with a dramatic article in Wired magazine.

A Wired journalist took a spin in a hacked car, which the researchers remotely drove into a ditch [49]. This announcement created waves both in the automotive industry and among the general public, and continues to inspire both continued media and public discussion, as well as safety legislation. The hack led to a recall of 1.4 million cars, the proposal of new vehicle cyber safety regulation in Congress, and a $400 million drop in Fiat Chrysler's market cap [51].

This incident highlights the deficiencies of the automotive industry with regards to safety, and the adverse effects of informal software engineering methodology on both end consumers and the bottom line. Fiat was lucky in that the two security researchers chose to disclose this exploit. More exploits along the same vein are sure to exist. How then can runtime verification technology help automobile manufacturers improve vehicular safety?

In this specific case, RV-ECU could have come into play in multiple ways. The researchers mention in their Blackhat conference paper that, to their surprise, while the Jeep's firmware update mechanism was designed to be operated via the dashboard display, nothing prevented them from sending firmware update commands over the air, without authentication. This entire attack approach would have been rendered invalid with one simple global safety property formalizing the requirement that firmware updates must be driven from the dashboard display only.

This does not, however, deal with the more fundamental problem that CAN traffic is unauthenticated and multicast. This means that all an attacker needs to do to gain control over an automobile is gain access to the CAN bus and impersonate legitimate

ECUs. Through local and global monitors, RV-ECU easily allows the implementation of authentication and authorization protocols as lightweight formalisms completely orthogonal to the functionality of the software components. In other words, the engineers developing the code that achieves the desired functionality of the ECU need not worry about authentication, that being added automatically by RV-ECU. This achieves a separation of concerns that makes authentication and authorization simpler and more portable.

Even if the researchers found a way past that, proper formalization of vehicle safety would prevent many of their attacks from taking place, even if they could impersonate legitimate ECUs. We cannot assume that the automotive industry will be able to correct all security vulnerabilities that could lead through compromise through traditional testing and analysis: even in the payments industry, where security has been a key focus and source of spending and concern, recent studies have concluded that the complexity of modern software systems makes breaches virtually impossible to avoid [52]. Such a conclusion likely also applies to automotive, with increasingly connected and complex systems implying that the elimination of all security-sensitive software errors and user error is unlikely if not impossible. We must thus manage the risks entailed by a compromise, providing a trusted hardware base that is minimal and well verified to ensure the integrity of the global system regardless of any malicious actions taken by the attacker.

Even if a relevant specification were not preinstalled with the vehicle, new safety specifications could cause the vehicle to be updated with the specification at a later date and protect all newly sold vehicles from exhibiting the same problem. With no impact on functionality assuming correct operation of the specification, the costs to test, implement, and distribute the safety updates would be significantly less than that of a dealership-based reflash of the entire ECU, a change directly affecting both the safety and the functionality of the component.

Beyond ensuring the enforcement of functional properties despite a security breach, RV-ECU can also protect the system from a malfunction, helping to curtail automotive recalls. Figure 6 shows an analysis of past automotive recalls. We look only at recalls occurring in the last five years and affecting more than 50k cars, with software errors as the principal contributing factor to the recall.

The results of our initial analysis seem quite promising: simple, English-language properties that are portable across vehicles and manufacturers are often enough to have entirely prevented the recall assuming the presence of a functional runtime verification platform. Of all the recalls we analyzed, only two were not preventable by runtime verification: in both cases, the error causing the recall was a mistake in the specification of the original system rather than in its implementation, meaning that the RV-ECU would potentially enforce incorrect behavior and would not improve or alter the safety of the overall system.

Unlike other implemented and practical systems, our formalisms are quite concise. Figure 7 shows one property featured in Figure 6, namely that the cruise control motor cannot send messages unless the cruise control is operated (started since last stopped). As you can see from the property, a simple regular expression over the cruise control messages and associated recovery action is sufficient to enforce the relevant property in our simulation. Our CAN API provides a standard for translating CAN messages into events, over which the regular expression above is written. We also provide a function to send messages on the CAN bus, with the component and payload defined as enumerations. Properties can be tested on a PC-based simulation or on-device, as long as specifications for the manufacturer-specific components of the CAN bus are provided. We have reverse engineered several such components on a 2012 Honda Odyssey.

FIGURE 6 Selected large software recalls since 2010 along with their preventability from monitoring. * = Authentication layer also required, ** = Physical component involved, property may not be sufficient.

Automaker	Type	Year	# Vehicles	Relevant Property	Monitoring
Toyota [53]	Mechanical	2010	7.5M	Acceleration messages cannot overlap braking messages on the bus	CAN Only
Chrysler [54]	Software	2015	1.4M	Only messages in scope of the target ECU should be processed	CAN+Local*
Toyota [55]	Software	2015	625K	While moving, the hybrid system can only be shut down through the ignition switch**	CAN Only
Ford [56]	Software	2014	595K	The airbags must deploy within 10 milliseconds of acceleration over a threshold on any axis	Local Only
Ford [57]	Software	2015	432K	The engine cannot be running without the key in the ignition	CAN Only
Honda [58]	Software	2015	92K	Not preventable via RV (specification error)	Local Only
GM [59]	Software	2014	52K	Not preventable via RV (specification error)	Local Only
Jaguar [60]	Software	2011	18K	Cruise control motor cannot send control messages unless cruise control has been started since last stopped	CAN Only

FIGURE 7 The ERE-based safety property enforcing cruise control messages only while in scope.

```
ere : (cruise_control_start cruise_control_message* cruise_control_stop)*

@ fail {
  CAN_DO(CruiseControl, Stop, 1);
}
```

A Practical Demonstration

The first step towards demonstrating the separation of functionality and safety on a vehicle architecture using RV-ECU is the creation of a real-vehicle demo showcasing our architecture monitoring a realistic but simplified safety property.

Consider the following body-related property of door safety in a minivan, with electronic controllers that open the rear sliding doors in response to messages over the CAN bus: Unless the driver has unlocked the rear doors from the global vehicle lock/unlock controls, and the doors have not been locked since, the motor responsible for opening the doors should not do so. The safety monitoring code of this property as well as its automatic generation using the technology underlying RV-ECU have been illustrated in Figure 1.

It is not difficult to imagine a situation in which this property could be violated. For example, with a malicious attacker gaining control of only the infotainment system, connected to the body CAN bus, the malicious attacker could easily spoof a "rear door open" message while the vehicle is moving at high speeds to endanger the safety of any potential rear passengers. Alternatively, even in situations where no malicious attacker is present, a malfunctioning ECU connected to the body bus anywhere in the car could create such an unsafe situation by sending a message to the motor to engage. Finally and most likely, a passenger seating in the rear seat may (mistakenly) push the door open button, which subsequently sends the motor engage message. The last scenario above is obviously checked by almost all cars, likely using a protocol implemented in the door ECU that sends data-collecting messages to other ECUs and then sends the motor engage message only if it determines it is safe to do so. Not only is the door ECU more complex than needs to be due to mixing functionality and safety, but the overall systems is still unsafe, because the other two scenarios can still happen. With RVECU, all three scenarios above are treated the same way, with a global monitor ECU in charge of monitoring the safety property possibly among tens or hundreds of other similar properties, and with any other ECU free of developer-provided safety checking code.

The overall system is simpler and safer.

We have obtained a STM3210C-EVAL development board implementing the popular STM32 embedded architecture. We are mimicking a minimal AUTOSAR-like API exclusively for interacting with the CAN bus, and running our certifiable high-assurance code to monitor and enforce the previously mentioned property in a 2012 Honda Odyssey minivan. Our demo is implemented and working, and we intend to demonstrate it as part of our presentation in SAE 2016. Figure 8 shows our development embedded board running on the CAN bus of our demo vehicle, attached through a connection in the driver's side lock control unit. Figure 9 shows the FSM-based property we monitor in our initial demonstration of the body CAN, available at https://runtimeverification.com/ecu. In English, the property states that the headlights should be on whenever the windshield wipers are on, and set to the user's selected mode when the wipers are off. While this is likely not an entirely realistic property (as manufacturers wish to grant users the ultimate control over headlight state), it serves as a good demonstration for the ability of our monitoring platform to enforce real properties on the global CAN bus.

Despite the simplicity of this formalism, the need to maintain regularity imposed by using an FSM render the property quite verbose. As you can see, for a simple one-line English property, the property monitored in a vehicle is over 20 lines in length. Still, the property is relatively simple to understand and create: we have one state for each possible (wiper, headlight) state pair, where the wipers and headlights can either be on or off. In this example, when we refer to headlights we are referring to the standard night time low beams of our test vehicle. We then have one transition from each state for each possible change in state of the subcomponents. For example, if the

FIGURE 8 RV-ECU development prototype connected to the body CAN bus of a 2012 Honda Odyssey.

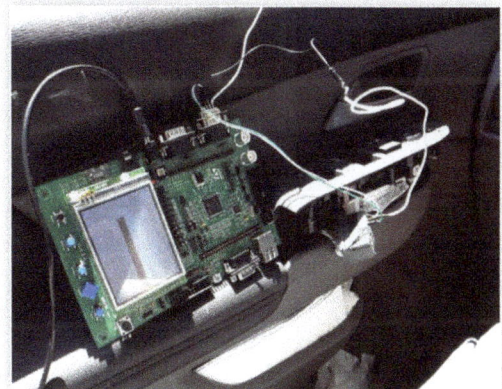

wipersOn event is seen in the

wipersOffHeadlightsOff state, we transition to the

wipersOnHeadlightsOff state.

Lastly, we have an unsafe state (

wipersOnHeadlightsOff), and a recovery handler (

@wipersOnHeadlightsOff) which sends a message to the CAN bus using our built-in CAN communication API to turn the Headlight component to High one time (

CAN_DO(Headlight, High, 1);) any time the associated state is entered.

The FSM-based safety property experimentally enforced on the 2012 Odyssey.

```
fsm :
 wipersOffHeadlightsOff [
   wipersOn -> wipersOnHeadlightsOff,
   wipersOff -> wipersOffHeadlightsOff,
   headlightsOn -> wipersOffHeadlightsOn,
   headlightsOff-> wipersOffHeadlightsOff
 ]
 wipersOffHeadlightsOn [
   wipersOn -> wipersOnHeadlightsOn,
   wipersOff -> wipersOffHeadlightsOn,
   headlightsOn -> wipersOffHeadlightsOn,
   headlightsOff-> wipersOffHeadlightsOff
 ]

 wipersOnHeadlightsOn [
   wipersOn -> wipersOnHeadlightsOn,
   wipersOff -> wipersOffHeadlightsOn,
   headlightsOn -> wipersOnHeadlightsOn,
   headlightsOff -> wipersOnHeadlightsOff
 ]
 wipersOnHeadlightsOff [
   wipersOn -> wipersOnHeadlightsOff,
   wipersOff -> wipersOffHeadlightsOff,
   headlightsOn -> wipersOnHeadlightsOn,
   headlightsOff -> wipersOnHeadlightsOff
 ]

 @wipersOnHeadlightsOff {
   CAN_DO(Headlight, High, 1);
 }
```

This recovery means that the unsafe state will never be the permanent state of the system. The transition out of the unsafe state is driven by a response from the monitor to the transition into the state, showing the possibility of recovering from property violations using only bus messages. On a real vehicle, the effect of running this monitoring code on our prototype RVECU which is connected to the CAN bus is that it is impossible to turn the wipers on without having the headlights turn on, regardless of the position of the headlight controls. User control of the headlights is also returned any time the wipers turn off, and the monitor enters a set of states it knows to be safe (as any states with wipers off are known to be safe).

This property can alternatively be specified in the more concise past time linear temporal logic (PTLTL) formalism, also supported by RV-Monitor. In this formalism, the formal definition of the property would be: [](wipersOn => (headlightsOff, headlightsOn]). This property states that it is always the case that when the wipers are on, the headlights have not been turned off since they have been turned on (using interval notation). While we do support this notation in RV-Monitor and this representation showcases our ability to concisely express formal specifications, we believe that past-time linear temporal logic is beyond the immediate familiarity level with formal properties of the majority of our target with the RV-ECU product. We will thus focus primarily on the familiar FSM and regular expression formalisms, common in general practice.

Future Work and Applications

One unanswered research question regarding the proposed RV-ECU safety architecture, shared with other formal analysis methods in the automotive domain, is what is the ideal formalism suited for mathematically defining automotive properties. Runtime Verification, Inc., will work with their automotive partners and customers to provide an intuitive domain-specific formal representation and associated plugin for our system allowing safety engineers or managers to comfortably specify such properties, lowering the barrier to entry for our technology and facilitating its uptake in industry. Such a plugin would likely also support the definition of real-time and temporal safety properties to fully specify the range of possible safety specifications associated with a safety-critical real time system.

As part of this process, we are seeking an automotive manufacturer or supplier willing to experiment with our technology in their development environment, evaluating the benefits of our specification language, code generation infrastructure, and the general separation of safety and functionality we provide to the specification and monitoring of complex software systems.

Technical Limitations and Drawbacks

There are several limitations and drawbacks raised by the potential vehicle architecture we propose. The first is the additional communications on the CAN bus required between the local and global monitors. In the proposed architecture, traffic can be as much as doubled for safety critical messages when the RV-ECU acts as a relay point, as compared to when safety is not checked at all. In an already overcrowded bus, such limitations could be prohibitive to the implementation of our solution. To mitigate this, it is possible to monitor properties primarily locally, monitoring only properties involving multiple ECUs through the global safety monitor. It is worth noting that existing architectures also require a number of messages to be sent specifically for checking safety, e.g., the door ECU requesting data from other ECUs when the door open button is pushed; more research is needed to compare the number of messages that RV-ECU requires versus the existing architectures. In the long term, our hope is that efforts aiming to replace the CAN bus with faster and higher-throughput communication standards will allow for our additional communication without overburdening the system.

Another key technical challenge for our technology is developing a formalism and infrastructure capable of handling the real-time properties required by the automotive industry. Because we have no access to the proprietary specifications currently used, we are unable to develop such a system. We thus wish to start with an architecture capable of handling non-real-time properties, extending it with real time support as is required to handle the needs of our customers.

The main risk in the adoption and development of run-time verification in automotive however lies in the development of accurate, rigorous specifications which the automotive industry does not currently have in the development process. With only a vague, often informal notion of formal system safety, the majority of OEMs and suppliers have not fully and rigorously defined precisely what the safety of a vehicle system consists of. This initial effort to formalize the notion of safety in the vehicle may be cost prohibitive and difficult, but remains necessary for the eventual creation of a system with strong safety guarantees and high assurance. We believe this undertaking will have a positive effect on the automotive industry, providing a rigorous notion and understanding of what safety means in the context of the vehicle system. This rigorous notion will help at every level of the development cycle, facilitating testing, development of new functionality, and regulatory certification.

One further and clear technical limitation of our approach is its inability to protect from hardware faults. Because our approach operates at the software level, any flaws in the CAN driver being used or the hardware of any individual ECU can still cause problems undetectable and unforeseen by the specification monitoring system. While the former can be mitigated by full verification of the CAN driver, a more traditional fault detection approach is likely more suitable for detecting faults in the actuators, sensors, and processing hardware involved in the vehicle system.

It is also important to note that extensive full-vehicle testing will still be required despite the presence of our safety architecture. The effects of our monitoring code and the effects of the interactions of the specification monitors with the full system cannot be determined without testing. We hope that with rigorous, checkable specifications and descriptive error conditions, our system will speed the testing cycle for safety requirements by allowing rapid evaluation of the system against its stated requirements. Despite this, rigorous conventional testing is still required to maintain the safety of the full vehicle system.

Lastly, there is a risk that our specifications will themselves introduce safety risks in the system: if the specifications are inaccurate, unforeseen circumstances can create unexpected programmatic behaviors actually detrimental to the safety of the system. For example, in Figure 4, a monitor could theoretically override or exclude a message by the controller it incorrectly believes to be unsafe, which would itself cause safety problems in the vehicle. While this is undoubtedly possible, our belief is that any unforeseen behaviors in the formal specifications provided could just as readily be present in the code itself, which implements informal specifications. Formalizing the specifications implicit in the current codebase rigorously will not inherently introduce unforeseen behaviors, and we expect that such formal rigor in the testing phase will actually help reveal previously unconsidered safety-critical interactions. In the cases where there are unintended interactions between the monitors and the system itself, traditional testing should be able to reveal them at least as readily as it reveals inconsistencies between actual and expected behavior in current systems.

Conclusion

Thus, we claim that separating safety and functionality in the modern automobile system would help find software bugs early in development, avoid recalls, and improve communication between original equipment manufacturers and their suppliers. We propose runtime verification as one solution allowing for this separation, and introduce a potential architecture for realizing such a practical separation.

We see that specifications checked at runtime can be both concise and formally precise, allowing for their development by engineers and managers not trained in formal methods while ensuring they are modular and easily sharable. We implement such a system with a practical demonstration of a simplified body safety property, and lay out the roadmap for future work enabled by the separation of safety and functionality. We discuss the technical limitations and drawbacks of our approach, including resource overhead, incomplete specifications, and an inability to deal with low level hardware faults.

Overall, we seek to develop a commercial product usable by the automotive industry to add runtime verification to vehicles, both in the testing and development cycles and in production. We have already created a production-ready architecture for the provably correct monitoring of safety properties on automotive buses, and intend to partner with interested parties towards the application of such a system. We would like to apply such a technology to large-scale projects to analyze scaling concerns and demonstrate the feasibility of our approach in production, improving the overall safety of automotive systems.

Acknowledgements

We would like to thank our partners at Toyota Info Technology Center, Inc., and DENSO International America, Inc., for their collaboration, industry insight, and generous funding. We would also like to thank the National Science Foundation for funding work related to this project under SBIR (small business development) grants, specifically related to the analysis of past recalls, development of a prototype, and formal verification of runtime monitors. Additionally, we would like to thank NASA for SBIR funding that has helped improve the infrastructure of our code prover.

Further assistance for this work was provided by research and development performed at the Formal Systems Laboratory of the University of Illinois at Urbana-Champaign, including funding from NSF, NASA, DARPA, NSA, and Boeing grants.

References

1. Charette, R.N., "This Car Runs on Code," *IEEE Spectrum* 46, no. 3 (2009): 3.

2. Shinichi, S. and Mutsumi, A., "Automotive System Development Based on Collaborative Modeling Using Multiple ADLs," *ESEC/FSE 2011 (Industial Track)*, 2011.

3. Vellequette Larry, P., "Fiat Chrysler Recalls 1.4 Million Vehicles to Install Anti-Hacking Software," http://www.autonews.com/article/20150724/OEM11/150729921/fiat-chryslerrecalls-1.4-million-vehicles-to-install-anti-hacking.

4. Christiaan, H., "VW Ordered to Recall 2.4 Million Cars in Germany with Cheat Software," http://www.autonews.com/article/20151015/COPY01/310159986/vw-ordered-to-recall-2-4-million-cars-in-germanywith-cheat-software.

5. Reuters, "Toyota Recalls 625,000 Cars over Software Malfunction," http://www.dw.com/en/toyota-recalls-625000-cars-over-software-malfunction/a-18585121.

6. Associated Press, "Ford Recalls 432,000 Cars over Software Problem," http://www.dailyfinance.com/2015/07/02/ford-recalls-cars-software-problem/.

7. Associated Press, "Honda Recalling 143,000 Civics, Fits to Fix Faulty Software," http://bigstory.ap.org/article/2f5f75fd91e64ec6bde06bacf9824867/hondarecalling-143000-civics-fits-fix-faulty-software.

8. Eric,B., "GM Recalls Nearly 52,000 SUVs for Inaccurate Fuel Gauge," http://www.reuters.com/article/2014/05/03/us-gm-recallsuv-idUSBREA4209C20140503.

9. Ben, K., "Chrysler Recalls 18,092 Fiat 500L Cars for Transmission Issue," http://www.reuters.com/article/2014/03/17/us-chrysler-usrecall-idUSBREA2G0PU20140317.

10. Shinichi, S., Veena, M., and Hemalatha, M., "Test Suites for Benchmarks of Static Analysis Tools," *ISSRE 15 Industry Track*, NIST, 2015.

11. Christel, B. and Joost-Pieter, K., *Principles of Model Checking* (Cambridge: MIT Press, 2008), Vol. 26202649.

12. Dawson, E. and Madanlal, M., "Static Analysis Versus Software Model Checking for Bug Finding," *VMCAI*, Springer, 2004, 191-210.

13. Java Path Finder, accessed 2014-05-17, http://babelfish.arc.nasa.gov/trac/jpf.

14. NuSMV Home Page, accessed 2014-05-17, http://nusmv.fbk.eu/.

15. UPPAAL: Academic Home, accessed 2014-05-17, http://www.uppaal.org/.

16. Klaus, H. and Grigore, R., "Preface: Volume 55, Issue 2," *Electronic Notes in Theoretical Computer Science* 55, no. 2 (2001): 287-288.

17. Allen, E.E. and Clarke Edmund, M.,. "Characterizing Correctness Properties of Parallel Programs Using Fixpoints," *Proceedings of the 7th Colloquium on Automata, Languages and Programming*, Springer-Verlag, 1980, 169-181, ISBN: 3-540-10003-2.

18. Klaus, H. and Grigore, R., "Monitoring Programs Using Rewriting," *16th IEEE International Conference on Automated Software Engineering (ASE 2001)*, Coronado Island, San Diego, CA, USA, November 26-29, 2001, IEEE Computer Society, 135-143, ISBN: 0-7695-1426-X, doi:10.1109/ASE.2001.989799.

19. Amir, P. and Aleksandr, Z., "PSL Model Checking and Run-Time Verification Via Testers," *FM 2006: Formal Methods, 14th International Symposium on Formal Methods*, Hamilton, Canada, August 21-27, 2006, Jayadev, M., Tobias, N., and Emil, S. Eds., Vol. 4085, Lecture Notes in Computer Science, Springer, 573-586, ISBN: 3-540-37215-6, doi:10.1007/1181304038.

20. Andreas, B., Martin, L., and Christian, S., "The Good, the Bad, and the Ugly, But How Ugly Is Ugly?," *Runtime Verification, 7th International Workshop, RV 2007*, Vancouver, Canada, March 13, 2007, Revised Selected Papers, Oleg, S. and Serdar, T. Eds., Vol. 4839, Lecture Notes in Computer Science. Springer, 126-138, ISBN: 978-3-540-77394-8, doi:10.1007/978-3-540-77395-5_11.

21. Howard, B., Klaus, H., Rydeheard David, E., and Alex, G., "Rule Systems for Runtime Verification: A Short Tutorial," *Run-Time Verification, 9th International Workshop, RV 2009*, Grenoble, France, June 26-28, 2009, Selected Papers, 1-24, doi:10.1007/978-3-642-04694-0_1.

22. Feng, C. and Grigore, R., "Parametric Trace Slicing and Monitoring," *Tools and Algorithms for the Construction and Analysis of Systems, 15th International Conference, TACAS 2009, Held as Part of the Joint European Conferences on Theory and Practice of Software, ETAPS 2009*, York, UK, March 22-29, 2009, Stefan, K. and Anna, P. Eds., Vol. 5505, Lecture Notes in Computer Science, Springer, 246-261, ISBN: 978-3-642-00767-5, doi:10.1007/978-3-642-00768-2_23.

23. Howard, B. and Klaus, H., "Internal versus External DSLs for Trace Analysis -(Extended Abstract)," *Runtime Verification - Second International Conference, RV 2011*, San Francisco, CA, USA, September 27-30, 2011, Revised Selected Papers, 1-3, doi:10.1007/978-3-642-29860-8_1.

24. Moonjoo, K., Lee, I., Usa, S., Jangwoo, S. et al., "Monitoring, Checking, and Steering of Real-Time Systems," *Electr. Notes Theor. Comput. Sci.* 70, no. 4 (2002): 95-111. doi: 10.1016/S1571-0661(04)80579-6.

25. Eric, B., Hendren Laurie, J., Patrick, L., Ondrej, L. et al., "Collaborative Runtime Verification with Trace-Matches," *Runtime Verification, 7th International Workshop, RV 2007*, Vancouver, Canada, March 13, 2007, Revised Selected Papers, Oleg, S. and Serdar, T. Eds., Vol. 4839, Lecture Notes in Computer Science, Springer, 22-37, ISBN: 978-3-540-77394-8, doi:10.1007/978-3-540-77395-5_3.

26. Feng, C. and Grigore, R., "Mop: An Efficient and Generic Runtime Verification Framework," *Proceedings of the 22nd Annual ACM SIGPLAN Conference on Object-Oriented Programming, Systems, Languages, and Applications, OOPSLA 2007*, Montreal, Quebec, Canada, October 21-25, 2007, Gabriel Richard, P., Bacon David, F., Videira, L.C., and Steele Guy, L., ACM, 569-588, ISBN: 978-1-59593-786-5, doi:10.1145/1297027.1297069.

27. Volker, S. and Eric, B., "Temporal Assertions Using AspectJ," *Electr. Notes Theor. Comput. Sci.* 144, no. 4 (2006):109-124, doi:10.1016/j.entcs.2006.02.007.

28. Justin, S., Ketan, D., Xiaowan, H., Radu, G. et al., "Aspect-Oriented Instrumentation with GCC," *Runtime Verification - First International Conference, RV 2010*, St. Julians, Malta, November 1-4, 2010, Proceedings 2010, 405-420, doi:10.1007/978-3-642-16612-9_31.

29. Mads, D., Bart, J., Andreas, L., and Frank, P., "Security Monitor Inlining for Multithreaded Java," *Genoa: Proceedings of the 23rd European Conference on ECOOP - Object-Oriented Programming*, 2009, 546-569, ISBN: 978-3-642-03012-3.

30. Irem, A., Mads, D., and Dilian, G., "Provably Correct Runtime Monitoring," *FM '08: Proceedings of the 15th int. symposium on Formal Methods*, Turku, Finland, 2008, 262-277, ISBN: 978-3-540-68235-6.

31. Úlfar, E. and Schneider Fred, B., "SASI Enforcement of Security Policies: A Retrospective," *NSPW '99: Workshop on New Security Paradigms*, 2000, 87-95, ISBN: 1-58113-149-6.

32. Horatiu, C., Pierre-Etienne, M., and de Santana, O.A., "Rewrite Based Specification of Access Control Policies," *Electron. Notes Theor. Comput. Sci.* 234 (2009): 37-54, ISSN: 1571-0661, doi:10.1016/j.entcs.2009.02.071.

33. de Santana, O.A., Wang, E.K, Kirchner, C., and Kirchner, H., "Weaving Rewrite-Based Access Control Policies," *FMSE'07: Proceedings of the ACM Workshop on Formal Methods in Security Engineering*, 2007, 71-80.

34. Falcone, Y., Currea, S., and Jaber, M., Runtime Verification and Enforcement for Android Applications with RVDroid, *Runtime Verification* (Springer. 2013), 88-95.

35. "ProofObjects: Working with Explicit Evidence in Coq," accessed 2015-09-1, http://www.cs.cornell.edu/~clarkson/courses/sjtu/2014su/terse/ProofObjects.html.

36. "How I Discovered CCS Injection Vulnerability (CVE-2014- 0224)," accessed 2014-05-15, http://ccsinjection.lepidum.co.jp/blog/2014-06-05/CCS-Injection-en/index.html.

37. "Formalizing 100 Theorems in Coq," accessed 2015-09-1, http://perso.ens-lyon.fr/jeanmarie.madiot/coq100/.

38. "CompCert—The CompCert C Compiler," accessed 2015-09-1, http://compcert.inria.fr/compcert-C.html.

39. Xavier, L., *The CompCert C Verified Compiler* (2012).

40. Robbert, K., Xavier, L., and Freek, W., Formal C semantics: CompCert and the C Standard, *Interactive Theorem Proving* (Springer, 2014), 543-548.

41. Rui, Z., Rong, M., Qi, Y., Chanjuan, L. et al., "Formal Verification of Fault-Tolerant and Recovery Mechanisms for Safe Node Sequence Protocol," *Advanced Information Networking and Applications (AINA), 2014 IEEE 28th International Conference on. IEEE*, 2014, 813-820.

42. Thomas, K., *Introduction to Formal Hardware Verification* (Springer Science & Business Media, 2013).

43. Diogo, B., Stefan, W., and Christof, F., "Automatically Tolerating Arbitrary Faults in Non-Malicious Settings," *Dependable Computing (LADC), 2013 Sixth Latin-American Symposium on. IEEE*, 2013, 114-123.

44. Aaron, K., "Runtime Monitoring for Safety-Critical Embedded Systems," 2015.

45. Lee, P., Alwyn, G., Robin, M., and Sebastian, N., Copilot: A Hard Real-Time Runtime Monitor, *Runtime Verification* (Springer, 2010), 345-359.

46. Petroni Nick, L. Jr., Timothy, F., Jesus, M., and Arbaugh William, A., "Copilot-a Coprocessor-Based Kernel Runtime Integrity Monitor," *USENIX Security Symposium*, San Diego, USA, 2004, 179-194.

47. Rodolfo, P., Patrick, M., Marco, C., and Grigore, R., "Bus-MOP: A Run-Time Monitoring Framework for PCI Peripherals," Tech. rep. Technical report, University of Illinois at Urbana-Champaign, 2008, http://netfiles.uiuc.edu/rpelliz2/www.

48. Rodolfo, P., Patrick, M., Marco, C., and Grigore, R., "Hardware Run-Time Monitoring for Dependable Cots-Based Real-Time Embedded Systems," *Real-Time Systems Symposium, 2008. IEEE*, 2008, 481-491.

49. Andy, G., "Hackers Remotely Kill a Jeep on the Highway With Me in It," http://www.wired.com/2015/07/hackers-remotely-kill-jeep-highway/.

50. Charlie, M. and Chris, V., "Remote Exploitation of an Unaltered Passenger Vehicle," accessed 2015-08-14, http://www.illmatics.com/Remote%20Car%20Hacking.pdf.

51. "Remote Exploitation of an Unaltered Passenger Vehicle," https://www.youtube.com/watch?v=OobLb1McxnI.

52. Douglas, S., "Are Large Scale Data Breaches Inevitable?" *Cyber Infrastructure Protection* 9 (2009).

53. National Highway Traffic Safety Administration, "Consumer Advisory: Toyota Owners Advised of Actions to Take Regarding Two Separate Recalls," 2010, http://www.nhtsa.gov/CA/02-02-2010.

54. National Highway Traffic Safety Administration, "Part 573 Safety Recall Report 15V-461," 2015, https://www-odi.nhtsa.dot.gov/acms/cs/jaxrs/download/doc/UCM483036/RCLRPT-15V461-9407.pdf.

55. National Highway Traffic Safety Administration, "RECALL Subject : Inverter Failure may cause Hybrid Vehicle to Stall – NHTSA," 2015, http://www-odi.nhtsa.dot.gov/owners/SearchResults?searchType=ID&targetCategory=R&searchCriteria.nhtsa_ids=15V449000.

56. National Highway Traffic Safety Administration, "RECALL Subject : Side-Curtain Rollover Air Bag Deployment Delay – NHTSA," 2014, http://www-odi.nhtsa.dot.gov/owners/SearchResults?refurl=email & searchType=ID&targetCategory=R&searchCriteria.nhtsa_ids=14V237.

57. Chicago Sun Times, "Ford Recalls 432,000 Cars Because of Software Problem," 2015, http://chicago.suntimes.com/business/7/71/740316/ford-recalls-software-problem.

58. National Highway Traffic Safety Administration, "RECALL Subject: Incorrect Yaw Rate/FMVSS 126—NHTSA,". 2015, http://wwwodi.nhtsa.dot.gov/owners/SearchResults.action?searchType=ID&targetCategory=R&searchCriteria.nhtsa_ids=13V157.

59. National Highway Traffic Safety Administration, "RECALL Subject: Inaccurate Fuel Gauge Reading—NHTSA," 2014, http://www-odi.nhtsa.dot.gov/owners/SearchResults?searchType=ID&targetCategory=R&searchCriteria.nhtsa_ids=14V223000.

60. BBC News, "Jaguar Recalls 18,000 Cars Over 'Faulty' Cruise Control," 2011, http://www.bbc.com/news/business-15410253.

61. Oleg, S. and Serdar, T. Eds., "Run-Time Verification," *7th International Workshop, RV 2007*, Vancouver, Canada, March 13, 2007, Revised Selected Papers, Vol. 4839. Lecture Notes in Computer Science, Springer, ISBN: 978-3-540-77394-8.

epilogue

Juan R. Pimentel
Professor of Computer Engineering
Kettering University

In this book, we have characterized SOTIF, one of the main categories of automated vehicle safety. It is the safety category that is not due to component or element faults but rather to performance limitations or functional insufficiencies. Perhaps the best examples of SOTIF hazards are those due to functional insufficiencies of the perception system of an automated vehicle. There are a number of outstanding safety issues discussed in the book, which will hopefully help clarify this topic, such as the nature of SOTIF, its underlying sources, and specific risk reduction actions. There are two primary approaches to reduce risk, a management approach and a technical approach. The former is based on using system engineering methodologies, particularly the V-model, while the latter involves specific risk reduction measures or mechanisms such as redundancy. In addition to characterizing the safety category of SOTIF, we have included a set of ten papers that are representative of research in this category. Although there is a fair amount of work in this area, much more work is needed to solve a number of outstanding issues, for example, how to properly validate safety targets and how to deal with the inherent uncertainty and complexity of the environment and the intrinsic uncertainty of functional implementations of machine learning.

Accordingly, most SOTIF papers deal with using system engineering techniques or risk reduction measures. Reducing SOTIF risk is challenging because of the nondeterministic and uncertain nature of the environment, particularly during bad weather. In addition, the intrinsic uncertainty of the implementation of some data processing solutions, for example, machine learning, contributes to the challenging nature of validating automated vehicle safety. Papers that promote using system engineering techniques emphasize a thorough approach to generate safety requirement specifications together with detailed plans for testing, verification, and validation. Papers that promote using risk reduction measures or mechanisms emphasize error and failure detection together with fault-tolerant techniques aimed to achieve fault-safe, fail-operational, or fail-silent systems. In addition, the introductory chapter presents a risk-based model that integrates SOTIF into a risk-based approach. This requires additional definitions and/or more general interpretations of those in ISO 26262. The proposed model introduces a new source of error, *performance error*, that models accident sources in the SOTIF category. The topic of safety measures for measuring the level of safety is crucial for automated vehicle safety. One type of safety measurement is around standards, processes, procedures, and design, and it is most useful at the design and simulation stages. One example of a safety measure that can be used during design is the number of expected catastrophic failures per year of operation (e.g., 500 hours) [24]. Another example of a safety measure is MDTA, the mean distance to accident (e.g., in km or miles) used in SAE paper #6. It can be argued that it would be difficult for all stakeholders to agree on a universal set of safety measures at this low level of abstraction.

www.ingramcontent.com/pod-product-compliance
Lightning Source LLC
Chambersburg PA
CBHW052337210326
41597CB00031B/5286